The Moogh

Guy Lane

The Moogh
Guy Lane

ISBN: 978-0-9944203-3-6

The first draft of The Moogh was written and edited exclusively on CityCats, the Brisbane River ferry service, between 20 March and 5 September 2015. Thanks to the captains and crew of this awesome service, and to the Brisbane River for the muse. Thanks also to David and Agnieszka for proofing early drafts.

Visit the official website: www.guylane.com

Ambulance Chaser

It is ten to nine on a brisk London morning, and Maggie Tarp makes her way to the corporate headquarters of the Fractious News Network. Outside the building, she sees cigarette butts lying on the ground next to the planter and she shakes her head, disapprovingly.

"That is so disgusting," she mutters as she swoops past.

Inside the building, she clips across the marble floor to the elevator and presses the up button. Glancing momentarily at her reflection in the shiny metal doors, she turns away and contemplates what she's seen.

Her long brunette hair rests lightly around her neck like a sleeping Persian cat. She touches her face with sufficient makeup to create a look at least three points more delightful than her birthright. She wears just enough foundation to hide the pale band of freckles that cross her nose. Those damned dots give away her humble origins, why can't they just disappear?

Maggie brings all these colors and textures to perfection by an impressive wardrobe. Today, she wears a silk blouse tucked into a gray pencil skirt, with stockings and high-heels to match.

Maggie used to look a bit bland, with freckles across her nose, until she saw the movie Bladerunner. Since then she styled herself on the character Rachel from Tyrell Corp.

"Do you like our owl?" she'd sometimes say, practicing the role. It is a fitting question because Maggie is wise beyond her years. A successful professional at just 26 years old. Who would have thought a Tarp could come so far?

She glances back into the mirrored surface to take a measure of the one noticeable imperfection. There is a scowl crossing her face. It's always like this when she is away from the Moogh. The Moogh Scowl, they call it.

It's even worse when she's in the city, like today. There's too much traffic belching greenhouse gases and too many people rushing around, smoking cigarettes and dropping the butts on the ground. She does her best to force a smile, but it's not easy. Sometimes the Moogh Scowl lasts for days.

The doors of the elevator slide open, and Maggie steps back to allow the people inside to egress. She recognizes an annoying man and is thankful that a swirl of hair largely hides her face.

She moves into the elevator and presses the button for level 12, the Executive Suite. The silver doors slide close but just before they meet, the toe of an unpolished shoe jams into the gap. Maggie steps back in surprise as a pair of hands with chewed fingernails grip the inside edges of the doors. The man makes a grunting noise as he struggles to prevent the doors closing. "Can I do it?" he says, forcing the silver doors part-way open.

Then to Maggie's dismay, the journalist, Perrin Speer, squeezes through the gap into the elevator. She squirms, feeling like a snow pea trapped in a pressure cooker with a chili pepper.

Perrin repeatedly hits the 'close door' button, observing that the floor-12 button is lit. He starts talking in his high-pitched voice. "Whoa. Swanky. Executive Suite," he hits the button for the sixth floor and glances around to see Maggie glaring at him. "Howdy, Cow Girl," he says, provocatively.

Perrin is mid-thirties. He dresses op-shop and never seems to get a proper shave. To most people he is an annoying asshole, but the newspaper executives love him because he knows how to write populist shit about tragedies. Perrin Speer sells a lot of Fractious News.

"Get a life," Maggie grumbles, feeling the scowl move bone deep.

"It is a cow, isn't it? You know, the *Mooooog*." He pronounces the word with the accent of a mooing cow.

"Grow a womb, Perrin!"

Perrin is persistent, irritating and lacking in social intelligence. It is a combination that makes him successful in his line of work because he's not afraid to ask inappropriate questions.

"It's a simple question, Maggie," he persists, "is the Moogh a cow?"

"No!" Maggie bites back. "The Moogh is not a cow! Nobody knows what the Moogh is. It's just a Moogh."

"Maybe it's a mutant cow and the G stands for GMO."

"Go to Mars, why don't you?"

The elevator arrives at the sixth floor. The doors slide open, but Perrin doesn't step out. Instead, he hits the 'close door' button again.

"I'll take you top," he says, saliently. The doors close, and Perrin looks at Maggie through the reflective surface. She has deliberately turned away, cursing her luck.

"I know something you don't know," he says, cryptically.

"You know how to chase ambulances."

The elevator halts at the 12th floor, and there is a long delay before the doors slide open. At last, Maggie can escape.

"Good luck with the Dim Director," says Perrin.

Maggie moves quickly into the Executive Suite, relieved to be out of the pressure cooker. She wonders for a second how Perrin knew that she was meeting with the Dim Director. It troubles her for a moment and then the thought passes.

At the reception desk, there sits a fittingly glamorous receptionist whose hair is a tangle of blonde curls. She smells of perfume and nail polish remover. Maggie introduces herself by showing off her gold-embossed business card.

"Would you take a seat over there?" The blonde receptionist indicates towards an arrangement of plush leather chairs around a glass coffee table.

Maggie takes a few hesitant steps in that direction,

conscious that she has been ordered to sit. Instead, she moves to the window and looks at the city below. Down, people are smoking and flicking their cigarette butts into the gutter. The grumpy look feels like it is settling in for the day.

Maggie doesn't know why she has been summoned to meet the Director, and she anxiously picks the edge of the card with a fingernail.

A familiar voice causes her to turn and see a girlfriend from her university days and now a co-worker at Fractious News. The two women perform a brief hug.

"It's the talented Maggie Tarp? says Novell, enthusiastically.

"Hi, Novell, how are you?"

"I'm fine, but look at you." Novell lifts the lapel of Maggie's blouse and teases the material between her fingertip and thumb. "Well, that's nice."

"That's caterpillar silk from a Catalan souk," Maggie says, bashfully.

"I'm so glad you're safe," Novell suddenly adopts a worried tone.

"Why would I be unsafe?"

"I heard that you work with that awful creature."

"I'm with the Moogh, and it's perfectly safe for women."

"Really? Didn't it rape an old lady, like King Kong or something?"

"What? Where did you hear that?" Public opinion of the Moogh is just awful in the offices of Fractious News, Maggie thinks, scratching the business card anxiously.

"It was trending in the blogosphere a while back."

Maggie groans. "Trending isn't true," she says for the millionth time. "The Moogh's a gentle giant. No one ever gets hurt."

"What about the guy with his leg missing? Or the one stabbed in the head?"

"But that wasn't the Moogh. That was--" Maggie starts

to plead, but she is interrupted by the curly haired secretary addressing her, telling her that she can go into the Dim Director's office.

"Let's have wine tonight," Novell says. "Do you have a card?"

Maggie passes over the dog-eared card that now has a whole corner ruined.

"Maggie Tarp," Novell nods approvingly, "Senior Moogh Reporter."

"Wine at seven," says Maggie. "And I'll set you straight about the Moogh."

The Dim Director

The Director's office is voluminous and dim, not unlike the man himself. He sits at a large redwood desk that is faintly illuminated by downlights. His face is cast in shadows as he peers at a document that contains a small number of short words printed in big letters. He holds a red pen in one hand and a rubber eraser in the other.

Maggie shuffles her feet, unsure whether he knows that she is standing right there. She feels anxious, wanting the meeting over with so that she can get back to the Moogh.

Eventually, the Dim Director raises his head. He rolls his eyes over her in the manner of a foodie appraising a new menu. It is as if he is pondering which item to sample first.

Maggie glances uncomfortably at the toes of her shoes, wishing that his hormones would just shrivel up and fall off forever.

The Dim Director's eyes are set too close together, and he has a small mouth. His jowls look either muscular or chubby depending on how the light falls on his face. He slowly moves his hands together, bringing the red pen and the eraser together. Then he interlaces the tips of his chubby fingers.

"Tell me something exciting," he says, eventually.

Taken off guard, Maggie's first instinct is to ask 'who me?' Instead, she says the second thing that comes to her mind. Some recent news about the Moogh. "Do you know the latest theory about how the Moogh is feeding itself?"

A blank look crosses the Dim Director's face. Maggie wonders whether she has said the right thing, and she waits patiently while her boss digests the question.

"Is it interesting?"

"Well, I think so."

"Let's start with that, then. See how you go."

"So you'll be familiar with the existing theory?" Maggie suggests.

"I am not a big Moogh fan, Maggie, so I'm not up to date," he says, waving a hand dismissively.

"Okay," Maggie thinks about how to frame the words so that the Dim Director can understand the science. "Well, the Moogh has never been observed to eat, so researchers reason that it is metabolizing reserves of fat."

"Metabo-liz-aling? What?" the Dim Director mispronounces. "That's not a word."

"Okay…"

"Anyway. Go on."

Off the hook, Maggie continues, "So the researchers calculated how much fat the Moogh would be burning, and they had soil scientists following the Moogh measuring the depth of its footprints. Before the Moogh Daisies grew, that is."

"Footprints?" asks the Dim Director, perking up.

"That's right."

"The Moogh has footprints? Now, that's interesting."

Maggie wonders how anyone smart enough to govern a global media corporation could fail to understand that an eleven-foot tall primate would leave footprints as it ambled across a field. She finds herself staring with her mouth open.

Recovering quickly, she says, "From the footprints, the scientists were able to estimate the weight of the Moogh and this allowed them to calculate whether it was getting lighter, as would be expected if it was burning off its fat."

"Well, is it?" asks the Dim Director.

"No. It's not," Maggie says, enthusiastically.

The Dim Director shakes his head, dismissively, "Scientists get it wrong all the time. That won't trend in the blogosphere."

"Well, that's right. The new theory is the exciting story."

"I thought that was the new theory," the Dim Director looks perplexed.

"No, that was the old theory, and it was disproven."

"So, now I'm really confused."

"The new theory is that the Moogh is photosynthesizing."

"Stop right there," snaps the Director waving his meaty hand in the air again. "You intelligent people do this all the time. It's so annoying."

"What did I do?"

"You use these big words so that you can look all fancy-pants in front of your intelligent mates. Well, I'm not one of them. Fractious News is for simple people. Use small words."

"I don't think there is a small word for photosynthesizing," says Maggie, despondently. "It's the way that plants get their food."

"So the Moogh is a plant, now?" the Dim Director cocks his head.

"No. The Moogh is an animal. But researchers have found a gold-colored zooxanthellae algae growing in its fur. It's a symbiotic relationship."

"Another big word! Shorten it. Make it simple. Maybe you'll trend."

Trending. Ugh! If she hears that awful concept again, she'll scream. "Algae," she says.

"So the Moogh is having a relationship with algae?" the Dim Director asks, raising an eyebrow. "How does that work?"

"It's a symbiotic relationship, like with coral."

The Dim Director is totally bamboozled, but hanging in there. "So the Moogh is a coral reef, now?"

"No," says Maggie, patiently, "but it has characteristics of a coral polyp."

"So let me get this right," the Dim Director adjusts the pen and the eraser on the table so that they are both pointing in the same direction. "The Moogh is a solar-powered, algae polyp thing?"

"Sort of," Maggie says, for the want of a better reply.

"I like that. Why can't you write about that?"

"I will, but there's more."

"Really?" The Dim Director leans forward, intrigued.

"It seems that the energy consumed by the Moogh is much greater than the energy that could be produced by regular algae."

"So where does that lead us?"

"One theory is that the Moogh's algae are much more efficient than any known to science."

"Now, that's interesting!"

"Absolutely, it is!"

"So the Moogh might be good for big solar panels?"

"Well, maybe…"

"I like that," says the Dim Director, excitedly, "Moogh saves the planet. Climate Change, just Moogh it. Global warming, get a Moogh on. There's an angle."

Maggie shows a look of deflated disappointment.

"You don't like that?"

"Why does the Moogh only have value when it is doing something for someone?" asks Maggie, frustrated. "Why can't we just seek to understand it. Empathize with it. The Moogh has intrinsic value."

"Intrinsic to who?"

"Intrinsic unto itself."

The Dim Director leans back in his seat and lets out a belly laugh. When he composes himself, he leans forward, seeming to implore Maggie to see his point of view.

"Capitalism," he says.

"What about it?"

"Does the Moogh consume a lot of media product?"

Maggie huffs, feeling trapped. "No. The Moogh doesn't read the news."

"And you work for a news outlet, right?"

"I do," she sighs.

"Now, if the Moogh shuffles across a field and stops to look at the clouds, that has no news value, given that's all it ever really does."

"It ambles and poses," growls Maggie, correcting the Dim Director's lousy Moogh English.

"But if the Moogh trips over and crushes some snotty-nosed kid to death on his birthday, that's front page material. Another Moogh death."

Maggie starts to irradiate. "There has never been a Moogh death!" she snaps. "The Moogh is entirely peaceful!"

"People get killed around it all the time."

"By other people!"

"Maggie, we call them Moogh Deaths so that the little people can understand. The public doesn't want to know about the Moogh; they want to know about the knifings and the beatings and murders that the Moogh incites. Then they spend money on Fractious News. Then we can pay people like you to, you know, write more news."

"Well it's just wrong," snaps Maggie, folding her arms across her chest.

"So here's the problem and why I brought you in." The Dim Director picks up the red pen and ominously moves it to the other side of the eraser. "Your personal views are at odds with the people who pay your wages."

"You are going to fire me?" Maggie asks, astonished.

"The Chairman wants me to. But I like you. I think that you're…" He sizes her up again. "Come up with a plan, they said. And if she comes good you can, you know… Keep her."

"Make it really clear to me," says Maggie, tersely. "I write thousands of words of copy for this network. I am embedded with moogh.org. I meet with their executives every day. I get inside the fence. What am I missing?"

"Not everybody loves the Moogh," says the Dim Director.

Maggie looks at the floor, gritting her teeth. The squinty-eyed dimwit is outsmarting her. How embarrassing is that?

"Some people hate the damn thing with a passion." The Dim Director lifts and then drops the pen on the desk for effect. "You know what's trending in the blogosphere right now?"

Maggie grits her teeth, trying not to shout, "F**k trending on the blogosphere!"

"You know what's trending right now? A Moogh story?"

"Tell me."

"There is a story about a US General, who wants to nuke the Moogh. Nuke it. With a bomb. A nuclear one. People are talking about that. Tweeting it. Liking it. Sharing it. Funny cat videos. All that stuff."

"So what do you want from me?"

"We need stories that bleed."

"Bleed," Maggie asks, grimacing.

"Action, conflict, drama. Turn your back on the Moogh and report on the sick and nasty things that people do to each other around it. Unleash the inner-mongrel, Maggie." The Dim Director gesticulates, trying to represent a hungry dog.

"What?" Maggie watches him, incredulously.

"But you don't have an inner-mongrel, do you?"

"What does that even mean?"

"Unleashing the inner-poodle just isn't going to cut it for the Chairman," the Dim Director says, despondently, shaking his head. "So you are a bit of a write-off, journalistically speaking. But you are inside the fence, you can get close to the Moogh. Maxine likes you, and that is valuable."

"What? Maxine?" Maggie is astounded that the Dim Director would refer to Maxine, the Chair of moogh.org, using just her first name.

"So I am not going to take you off the project. I am going to augment you."

"Augment me?"

"That's right. I am 'felici-tating' your 'augment-alation'

starting today." The Dim Director smiles, enjoying the words, even though they are big, and he doesn't know how to say them correctly.

"I don't understand."

"We are going to buddy you up with a real mongrel." The Dim Director grins broadly. He brings his hands together representing a mongrel buddying up with a poodle.

"Oh, no," Maggie sighs, anxious that she knows what's coming next. She watches as the Dim Director raises his hand, indicating for her to look around. She turns to see that Perrin Speer is in the room, sitting in an armchair with his legs crossed, listening intently.

"Howdy cowgirl," he says with a theatrical wave. He gets to his feet.

"No way!" Maggie blurts, instinctively. She feels her face flush with heat.

"He's coming with you," says the Dim Director.

"You can't do that!"

"It comes from upstairs."

"But he's a f**king space cadet," Maggie snaps.

"That's what everyone tells me," says the Dim Director.

"Excuse me, I am right here," says Perrin, stepping forward.

"Speer," says the Dim Director, firmly, "shut up!" He eyeballs Perrin long enough for the ambulance chaser to avert his eyes and look down at his feet, nodding.

Maggie presses for advantage. "They'll never let him into the Moogh Zone," she protests.

"Work with Perrin or find another job," says the Dim Director, firmly. "That's what the Chairman told me to say."

Maggie stares at the floor aghast, her mind churning. "I… I…" she stammers, but is unable to form a coherent sentence.

"I'll need your answer shortly," says the Dim Director.

He looks at his watch and says, "In the next ten seconds. Nine… Eight… Seven…"

Perrin Speer continues the countdown, "Six… Five… Four…"

White with Novell

A soft glow of orange light bathes the wine bar on the ground floor of the hotel where Maggie stays when she's in the City. A musician playing a grand piano enhances the ambiance. It is calm and peaceful.

Maggie sits at a small round table. In her hand is a small woodedn Moogh Mannikin. It's like one of those little figures that you can buy in art shops, except its in the shape of the Moogh. Maggine fondles this whenever she feels stressed, and twists the limbs into different Moogh poses. Today, she has twisted too hard, and she sits there holding the Moogh Mannikin on one hand and disconnected Moogh foot in the other. It is not a good portent.

She slids the mangled mannikin into her pruse, then clasps the stem of her wine glass and stares ahead blankly, stunned by the day's events. She has drunk half a glass of Chardonnay and already feels drunk and swoony.

Novell arrives, breaking her from her trance. They embrace warmly for a moment.

"You are one ahead of me," Novell nods towards Maggie's wine glass, "are you okay?"

"Not really," Maggie says feebly. She looks up at a waiter approaching.

"I'll have one of those," says Novell, indicating towards Maggie's glass.

"Me too," says Maggie.

"You look like you've had a shock."

"The Dim Director is forcing me to mix things that are unmixable."

"That sounds about right," Novell settles herself in her seat.

"I've been covering the Moogh for six years," says Maggie. "Right from the Genesis Amble on the first day. I saw it throw the Genesis Pose."

"I don't know what that means."

"When it first appeared. The Moogh. I was there on the first day. Just by chance, on a tourist bus. I know everything about it. I wrote the Dictionary of Moogh English. I know everyone around it. I'm hooked. A Moogh Junkie. I'm addicted, and I love it. I'm in it for enlightenment and to learn about something beautiful, wondrous and unique."

Novell nods her head gravely and says, "Go on."

"When I am around the Moogh, I am happy, and my life makes sense. But as soon as I am away, my life goes to shit. It's called the Moogh Blues, and I've got them right now. Big time."

"But you are going back, right?" Novell asks.

"Yeah. And typically the blues would be lifting by now. But I'll be accompanied by Perrin Speer."

"Oh, poor sister," Novell coos. She places her hand on Maggie's forearm.

"I have to take that knucklehead to my sacred place."

"There might be a good reason to do that," says Novell, sagely.

"Really?"

"Absolutely. There is a very sound reason to take something repugnant to a sacred place."

"What is that?" Maggie asks, hopefully.

The waiter returns with two glasses of wine and places them on the table while Maggie frets that her friend might lose her train of thought. Novell raises her glass and takes a long sniff of the wine. "Hmmm. That's nice?"

The delay seems endless. "What were you saying?" Maggie asks, anxiously.

Novell sips the wine, "Oh, that's delicious. What is it?"

"Come on, come on, come on," thinks Maggie, her energy draining away.

"Sacrifice," Novell says, eventually.

Maggie laughs aloud, realizing how anxious she'd

become. "There's a thought," she says. "I could slice him open with a glass knife."

"You'd win friends back here if you did that."

"But not with the Dim Director," Maggie looks at her lap, despondently. "Besides, the Moogh is not something that you sacrifice to. It doesn't have that energy."

Novell suddenly becomes agitated, like an insect had just bitten her. "How?" she says, and then seems lost for words.

"What?" asks Maggie, surprised by the burst of energy.

"I just got hooked on the Moogh. I need to know everything there is to know. Did you say there was an app? I need to download the app."

Novell retrieves her smartphone and swiftly googles the term 'moogh app'. "How close have you got to it?" she asks. "Is 'it' even the right word?"

"There it is." Maggie points to Novell's phone. "That's the Moogh App. The Adherents call it either 'he' or 'she' depending on the day. It sort of shifts. There are people researching that."

"You see, I don't even know what that means," Novell, is perplexed.

"I touched it once."

"I thought that was banned."

"It is, now."

"You touched it?"

"I was in its path, and I wasn't paying attention," Maggie's eyes light up as she rmembers. "It ambled right up to me, so close that I totally froze."

"Oh, my," Novell places her fingertips across her mouth.

"At the last moment," Maggie continues, "I reached out my hand and ran my finger across its belly."

"What did it feel like?"

"Soft. Its skin looks like leather, but it's actually fine hair, like on a sea lion."

"A sea lion?"

"Yeah, funny, huh?"

"And which finger?"

"This one," Maggie holds up the finger that touched the Moogh.

"Can I get a photo of that finger. A selfie with you in it?"

"Let's do it," Maggie grins and remains with her finger held in the air. Novell moves next to her and positions her smartphone. She photographs the pair of them with the finger in the foreground. Once the camera has clicked, she says, "That is so going on my profile page."

"That's so Mooghy," says Maggie, blushing.

"Of course, none of it makes any sense," says Novell, shaking her head. "I mean where did it even come from?"

"There's lots of theories. The one that I like is that it's ancient. It has been asleep, frozen or hibernating or something. And the global warming woke it up. It's like some messenger drawing our attention to what we humans are doing to the environment. That's what the Adherents believe."

"Adherents? You used that word twice now."

"That's what we call the people who move around with the Moogh. The permanent ones."

"How many are there?"

"There's about twenty thousand Adherents. And maybe a few hundred thousand flybys."

"Flybys?"

"Fly-by-nighters. Day trippers. People who come in for a little while and then leave."

"That's incredible," says Novell, taken by it all. "That's so many people."

"Yeah. A lot of individuals. A lot of catering. A lot of organizing. And you can't plan it because the Moogh gives no indication of where it is going next."

"It just wanders around?"

"It ambles. That's the proper Moogh English."

"And what's the story about the thing in its hand?"

"The Moogh's right fist is always clenched. It's believed that it is holding something inside?"

"Like what?

"No one knows. It's all just wild speculation."

"What do you think it is?"

"I spent many sleepless nights thinking about that. In the early days, we used to sit around till all hours, drinking Moogh Mead and having 'D' and 'M's about the Amble and Pose of Moogh Destiny."

"What?" Novell looks completely bamboozled.

"I think it is a…" Maggie seems to drift off, staring across the hotel lobby. When she comes back, she sees that Novell is staring, her mouth open in expectation.

"I think that it's a baby Moogh," Maggie says, at last.

Instantly, tears well in Novell's eyes and her face flushes red. "That makes me cry," she says, embarrassed. Maggie hands her the serviette from under her glass, and they both laugh as Novell dries her eyes.

"And to think that you have to take that muppet, Perrin Speer, over there."

"Yeah," says Maggie, raising her glass, glumly. "That makes me cry."

A Poster of Poses

Next day, Maggie rings the office and learns that her flight departs that afternoon and that she is to meet Perrin at the airport. She spends the morning listless in her hotel room, depressed. She calls Novell, hoping to revisit the conversation about sacrificing Perrin with a glass knife, but the phone rings out.

Finally, she drags her suitcase out of the hotel lobby and towards a waiting taxi. She is hoping that the cab driver will help her lift the bag into the boot without asking. However, as soon as he approaches, the driver asks, "Can I assist you with your bag?" and that ruins it completely. Maggie waves him away and loads the suitcase in the boot on her own.

At the airport, she finds Perrin in the departure lounge news agency, standing in front of the Moogh section. He is making annoying trombone noises with his mouth as he quickly flips through one of the Moogh magazines.

"Which one do you recommend?" he asks, when he glances around and sees her standing there, looking at him mistrustfully.

"They are all trite, populist shit," says Maggie, matter-of-factly.

"So which is the best?"

"You'd like them all equally," she says.

"I'll take the one with the glossiest cover, then."

Maggie watches as Perrin transacts for the magazine. He shoves the Moogh mag into his man bag, slung over his shoulder.

"Come on, we have to go," says Maggie, morosely.

"Which gate is it?" Perrin asks.

"You don't even know which gate it is?"

"I know it's one of the gates, that's a start."

"It's Gate 43." She moves away and Perrin trots along beside her.

"So," he says, "I need a crash course in Moogh, so I am an expert by the time we touch down."

"Well, that's not going to happen."

"What, you're not going to brief me?"

"I'll brief you, but I have been around the Moogh for years, and I consider myself a novice."

"Maybe you're not that bright," says Perrin, dismissively.

"I beg your pardon?" Maggie stops dead and stares at him, astounded. She shakes her head, disbelieving and continues on her way, upping her pace, hoping to leave Perrin behind. But the Ambulance Chaser walks as fast as he talks.

"I wouldn't be doing that if I were you," says Perrin, chuckling. "Begging me, that is."

"There's no technical manual for the Moogh," Maggie continues with a grumpy tone. "It is a phenomenon, never seen before, that calls into question every *a priori* assumption about what is genuine and honest and--"

"Yeah, yeah, yeah," Perrin interrupts, waving his hand in the air.

"Don't you shush me!" Maggie protests, angrily.

"You're missing the point somewhat, Paprika," says Perrin, "I'm not here to write about the Moogh."

"What? Paprika? What are you even talking about?"

Perrin starts sniggering uncontrollably. He puts his hand to his mouth to arrest his laughter. "It's okay. It's okay," he says, trying to change the topic.

"There is absolutely nothing good about any of this." Maggie shakes her head and starts mumbling to herself out of sheer frustration.

"Hey, Paprika!" Perrin calls out from behind her. She turns to see that he has stopped in his tracks and is pointing to a large wall poster.

"I mean like, shot the duck?" he says, with a questioning look.

Gate 43 is the route that many thousands of people take

to get from the City to the Moogh Zone. Decorations in this section of the airport include a variety of Moogh paraphernalia. One piece is a poster that shows simple line diagrams of the Moogh holding different poses. In one, it is looking upwards with its left hand pointing skyward. In another, it is leaning forward, like it was investigating something on the ground. There are nine poses in all and the diagrams look like a Tai Chi instruction manual.

"What has it been smoking, forever?" as Perrin, studying the images.

"They are Moogh Poses," says Maggie, flatly.

"And what does that even mean?"

"When the Moogh is not ambling, it is still. And when it is still, it holds its body in one of nine different positions, Moogh Poses."

"And what's all this going on?," Perrin waves his finger to indicate the illustration showing a flock of birds and bees and other insects flying around the Moogh and flowers around its feet.

"We call that the Entourage," Maggie tells him, patiently. "Wherever the Moogh goes, it is accompanied by many different types of birds and insects, including lots of big fat honey bees. The bees feed on the Daisies that grow in the Moogh's footprints."

"This is like Alice in Wonderland," Perrin mutters.

"That is the first thing that you have said that I agree with," Maggie says. She glances towards Gate 43 and for the first time in days, she gets a sense that she is escaping the City for the Moogh Zone. She looks back at Perrin and the crushing reality swoops back in. She's not going to the Moogh Zone alone. A thought crosses her mind. Maybe there is some way she can lose Perrin on the way. Now there's something worth contemplating.

Perrin glances her way, and she makes an innocent smile. They look at each other for a long beat. It seems for a moment that something new and exciting might flower from that first foray into agreeing with each other. The moment passes without anything grown.

Perrin breaks the brief silence by saying, “I wouldn’t get used to it.”

Maggie makes a snorting noise, “I will never get used to anything you do.” She turns and continues towards the departure gate.

No More Squashed Monkey

Maggie's bad mood deepens when she boards the plane and realizes that she has neither an aisle nor window seat, but the one in the middle. In the window seat is an old man with hair growing out of his ears and nose. He seems to have died, resting his head against the plastic shutter over the window. Perrin is in the aisle seat. Maggie dithers, knowing that the chair will hold her trapped for hours.

She takes her seat, grumpily, and immediately retrieves the in-flight magazine and stares at the pages, willing the time to pass faster. Perrin takes his seat next to her, making musical noises with his mouth and actively watching all the comings and goings on the plane. He swivels around, taking stock of everyone and making throwaway comments to people moving around him with no particular interest in the response he receives. Maggie observes this, sulkily, realizing with disappointment that for the most part, people seem to like the way Perrin behaves.

When the flight gets underway, Perrin retrieves the Moogh mag from his man-bag tucked under the seat in front of him. Maggie watches as he turns to a page with a double spread showing a close-up of the Moogh's face.

"Holy mother of pup!" Perrin exclaims theatrically. "What a fugly mutt! It looks like a squashed monkey!"

Maggie feels a flash of anger, and she audibly sighs in exasperation. She wants to shout, but she remains quiet as she thinks through an important idea. She knows that Perrin will continue to invent grotesque interpretations of the Moogh that will color every word that he writes unless she intervenes. She realizes that she can guide him towards an alternate view of the Moogh. She can shift his opinion before he starts sharing his awful thoughts with the world. To this end, she commits to engaging with Perrin – at least for as long as she can tolerate his annoying idiosyncrasies.

"Some people compare it a newborn," she says,

helpfully. "See the little furrows above the nose."

"A newborn what, exactly?"

"A child."

"And where's its wedding tackle?" Perrin asks, looking at a full-frontal photo of the Moogh.

"Its what?"

"Its fricking gonads, man."

"You're grotesque," says Maggie, recoiling.

"I'm grotesque? I mean look at this thing. It's like a Yeti dipped in fat."

"What!?"

"It looks like a snot gorilla."

"You're a freak!" snaps Maggie, furious.

"But I am creative, huh? You have to admit that, Paprika. And I ask the cutting questions that the public want answered. I'm serious, where's the willy or the, you know, the front bottom?"

Maggie grits her teeth. Then she gets back to her task of trying to shape Perrin's perception of the Moogh.

"The trending theory amongst the scientists is that it's like a bird. Its reproductive system only forms when it is mating season."

"I wouldn't mate with it. Would you?" Perrin holds an expression that looks like he was serious in asking that question.

"The Moogh's not my type," Maggie says and then feels conflicted for having contributed to his joke.

"And what's going on in this picture?" Perrin asks. The image shows a tent city and what looks like a hundred hippies dancing around a block of wood.

"That's a celebration on the day that the Declaration of the Adherents was signed," says Maggie.

"Well, I should have known that," says Perrin, sarcastically.

"Something you don't know?"

"What do I know? I do car crashes and family tragedies.

Give me a fire in a primary school I'll be all over it. I don't know Moogh shit."

"The Moogh doesn't shit," growls Maggie.

"Duly noted. So what's happening in this photo?"

"That photograph shows the day that the United Nations awarded the Moogh and the Adherents special status. It effectively told Governments to keep their hands off the Moogh and just let it amble. It was a huge day."

"Were you there?" Perrin asks.

"That's me right there." Maggie points to a patch of color in the middle of the crowd in the picture.

"You love the squashed monkey, don't you?" asks Perrin with an honest, open tone.

"Yes," Maggie says, plainly, like it was the first time she had been able to talk to Perrin without anger in her voice. "It's the best thing that has ever happened to me."

Maggie looks at Perrin and sees that he is shaking his head in wonderment, surprised by her answer. "Perrin, will you do something for me?"

"What?"

"Will you please not insult the Moogh in my presence? As a matter of professional and social courtesy."

Perrin bites his lower lip, nodding his head, quietly. They are interrupted by the air hostess commenting on the photos.

"Oh, that's the Moogh," she coos.

"I'm a Moogh reporter," says Perrin, quickly, putting on a voice. "My assistant Paprika and I are on our way to report on the magnificent beast right now."

"Well, that's so cool," says the air hostess, impressed.

"I'm writing a story about the Moogh's detachable penis," Perrin continues.

"Well, you're the right guy for that job," the hostess quips. "But I thought that the Moogh was female."

"Two-five-five, two-zero-two, zero," says Maggie, cryptically.

"She's giving me her phone number, now. It's the effect the Moogh has on women."

"That's the color code for Moogh Orange. The color of unspecified gender," Maggie explains.

Perrin addresses the air hostess, "Well, that's right. This is advanced Mooghology. You're with the experts now."

"Wow," says the hostess to Maggie. "Can I get you champagne?"

"Oh, thanks. Just a small one." Maggie blushes.

"I'll have a scotch on the rocks," says Perrin.

"Small champagne it is," the hostess says and then moves away.

Maggie grins and turns her attention back to the in-flight magazine. "*Touche*," she thinks. If she can 'touch' him, maybe she can win.

Moogh Happy Again

At some point, Perrin's nervous energy needs to be recharged, and that process begins in an instant. One minute his mouth is open and a torrent of quick-witted and insensitive words are pouring out, and next minute he is fast asleep. Unsurprisingly, Perrin is annoying even when he is not awake. He emits a rasping noise from his nose as he slumbers.

Maggie looks over at him, shaking her head wearily. She's wracking her brains, trying to figure out how to ditch him before they get inside the Moogh Zone. Maybe she can take advantage of his famed binge drinking, she thinks.

She notices his brown leather man-bag under the seat in front of him and she reaches down and pulls the bag onto her lap. While Perrin snores, Maggie investigates the bag's contents, forming a plan. She helps herself to some of his stuff, and once the bag is back in its place, she presses the overhead button and calls the air hostess.

Maggie orders two scotches with no ice, and when the drinks arrive, she nudges Perrin awake with her elbow. He comes to with a jolt and his mouth immediately opens.

"Let's go and pick a Moogher Booger," he mutters. "Are we there yet?"

"Here, drink this," Maggie, passes over a plastic cup containing both portions of whiskey.

"Whoa! I could take a liking to you, young miss, missy-miss," Perrin mutters as he takes the cup.

"Stop talking and drink up," Maggie instructs him, "Come on, skull, skull, skull."

"Yes, Ma'am." Perrin glugs the scotch back in one go. He grimaces and shakes his head like a wet dog.

Maggie takes the cup from Perrin's hand, "Go back to sleep."

"Yes, Ma'am," says Perrin and then promptly complies.

With Perrin back in the land of nod, Maggie relaxes,

feeling content that her plan is sound. As the flight continues, she calls on the air hostess two more times and doses Perrin with more scotch.

Perrin sleeps all the way through the touchdown, and it is only when the seatbelt sign makes the 'pong' noise that Maggie elbows him roughly on the arm. "Oy! Wake up," she snaps, and Perrin comes to with a jolt.

"Oh hell, what happened?" he clutches his head and moans.

"Listen up," says Maggie, firmly, "I've been through this airport before. It's busy and easy to get lost so follow my lead, alright?"

"Dominatrix, I like," says Perrin. He wears an expression of someone who is in pain.

"Are you okay, you look a bit woozy?"

"Bugger me, how much did I drink?" he puts a palm against his eyes, grimacing.

"You were going pretty hard."

Off the plane, Maggie moves quickly through the airport towards customs. She joins the shortest queue and directs Perrin to the longest line. He ambles along, drunk and disoriented, and clutching his head.

Maggie passes through customs and looks back to see Perrin approach the counter where a surly Spanish customs official repeats his only word of English, "Passport."

Perrin fumbles around inside his man bag but is unable to locate his documentation. The grumpy customs guy makes his request again. Perrin redoubles his efforts. He checks his bag again, pats his pants pockets and then goes back to the bag.

Maggie watches as a security guard moves in Perrin's direction. She chuckles to herself and then continues on her way. Outside the airport terminal, she finds a rubbish bin, and one by one drops Perrin's SIM card, passport, credit cards and driver's license into the bin.

Then she steps into a waiting taxi and instructs the driver

to take her to the Moogh Zone. Slumping back in the seat, she feels light and overwhelmingly happy. She can sense the muscles in her face relax, and her smile returns. After three days of Moogh Blues, Maggie is Moogh Happy again.

The Moogh App

Inside the taxi, Maggie retrieves her smartphone and opens the Moogh App. The first screen is a Mooglemap. It shows that the Moogh has ambled three and a half kilometers since she last logged in. Presently, it is located outside of a small village about five kilometers from the Mediterranean coast. The icon on the map shows that the Moogh is stationary and holding the #7 pose, the one they refer to as *Atten Hut*. It has been like this for five hours, so says the app.

The Moogh App also features a 3D model of the Moogh and nine buttons that activate the model to show the different Moogh Poses with their colloquial names:

Pose #1 - I've forgotten something

Pose #2 - My shoelace is undone [#2L= left foot] [#2R = right foot]

Pose #3 - My head is on fire

Pose #4 - Stargazer

Pose #5 - Snail's pace

Pose #6 - Dr. Good

Pose #7 - Atten-Hut! [#7L= left hand] [#7R = right hand]

Pose #8 - Swan Dive

Pose #9 - From whence it came

In one of the poses, the Moogh is standing, looking ahead with head tilted at 45 degrees above the horizon and with the fingers of its left hand against its mouth. This pose is referred to as 'I've forgotten something' because it causes many people to say, "it looks like it's forgotten something and is trying to remember what it is."

Another pose is called 'my shoelace is undone' for the obvious reason that the Moogh is looking down at one of its feet pushed forward. There are two variants of this pose, left foot forward, and right foot forward.

The pose that gets the most attention and makes it to the

front pages of newspapers around the world is called *my hair is on fire*. In this pose, the Moogh has both arms raised and its hand touching the top of its head.

There are poses called 'the Stargazer,' pointing into the sky, 'snails pace' bending forward and looking at the ground a few meters ahead. *Atten hut!* is where the Moogh has its clenched right hand pushed against the side of its head, and there are two variants, left hand and right hand. In the 'Swan Dive' pose, the Moogh has its arms swept back as it leans forward, looking at a point fifty meters in front. 'From whence it came' is where the Moogh is looking over its shoulder with its left arm cocked forward.

The Moog App also features a currency converter that allows anyone to use a credit card to buy Moogh Coin. Moogh Coin is the official currency within the Moogh Zone. It exists only online and can be transacted via smartphones. It is created using a cryptographic algorithm called 'proof of amble'. In simple terms, so long as the Moogh is alive and doing what it has done for the past six years, ambling and posing, the currency will continue to be created and tradable. However, if the Moogh were to die, then the Moogh Coin currency would collapse and become worthless.

Many people, even those who never visit the Moogh, buy Moogh Coin in order to gamble in the Moogh Casino. This is feature of the Moogh App through which users can engage in games of chance relating to the Moogh. It is possible to bet on a range of Moogh parameters including average speed of the Moogh's Amble, the chances of the Moogh throwing one of its nine poses and even the outcome of scientific research into the Moogh's origins and biology.

Moogh Coin was invented by Maxine Slivkin, the Chair of moogh.org, and very few people understand how it works. This fact does not deter hundreds of thousands of people around the world from buying them and gambling with them in the Moogh Casino.

For moogh.org, creating the online currency and selling it for cold, hard cash is an extremely profitable enterprise. It allows them to live in the top suites of luxury hotels as they conduct their governance duties. Proof of amble is also a clever mechanism to incentivize moogh.org to ensure the Moogh's safety and not milk it too hard. If the Moogh were to die, the money train would stop.

Another valuable feature of the Moogh App is the Moogh Messenger. This is used for communicating with the Councilors of moogh.org, the organization that coordinates all Moogh activities.

Using this feature, Maggie sends a message to Maxine, announcing her impending arrival. A few minutes pass and the messenger alert sounds. Maxine has recorded a video that runs for just a few seconds. She is Russian-English and she has a shiny, sun lounge-tanned skin. Blush and glasses with garish rims accentuate her features. Her eyes light up as she speaks into the camera.

"Well, hi Maggie," the smiling face on the video says. "Great to have you back. We've kept your room in the Plush Hotel. I'm running a presentation in the conference room. Come on down."

Maggie rests back in her seat, comfortable that her world is in order. She watches the scenery as the cab moves through the outskirts of a town, alert for the signs that she is getting close to the Moogh Zone.

The first evidence is the handiwork of some young miscreants from the Moogh Underground who have tagged the walls with Moografitti. Their artwork is recognizable for using a variant of the official Moogh Emoticon.

The official Moogh Emoticon shows a simple representation of the Moogh's face with vertical lines for its eyes. However, they are shown as circles in the Moogh Underground version of the emoticon. This 'eyes open' version is an exhortation to be wary. Not wary of the Moogh, but of the human agencies around it, particularly moogh.org and United Nations Moogh Affairs.

Years ago, Maggie had interviewed the leaders of Moogh Underground. She found them to be a fractious lot, undisciplined and with an inconsistent story. They told her of rampant corruption in the governing bodies including multimillion dollar deals signed without proper oversight. However, they offered no documentary evidence of these things.

Maggie sent this story to the Fractious News Network and was promptly offered a job as a full-time writer. This was back in the day when she wrote under the name of her alter-ego, Marequa Taard. Writing under her own name was one of the conditions of her employment with the Fractious News Network.

That was also when Maggie was taken under Maxine's wing and given access to moogh.org's inner circle. There were many privileges that came from the role as Moogh Reporter for Fractious News, embedded with moogh.org. There were luxury hotel suites, regular banquets and unfettered access to the Moogh. Maggie had thought little of Moogh Underground and their wild claims since she accepted this new life.

Maggie is distracted from her thoughts by a pattering sound as the cab drives through a swarm of fat honey bees. Dozens of the bees splatter across the windscreen, the glass smeared with nectar and bright orange clumps of pollen. The driver flicks the windscreen wipers and the water and the blades swat the dead bees away. He glances over his shoulder and says, with a thick Catalan accent, "Moogh Bees."

This is good news. The Moogh Bees signify that the Moogh has passed this way and that the Moogh Zone is close.

Maggie glances out the side window and observes

another sign that the Moogh Zone is close. There is a pitched battle being waged on a street corner between a crowd of young men hurling bottles and rocks at a phalanx of Level-4 Moogh Security. These foxy, female Level-4 wear white cat suits and respond by zapping young males with electric stun guns. One of the boys crawls along the pavement, grimacing, blood on his face.

That's not Moogh business, Maggie thinks, as she looks away. She turns her attention to the road block up ahead. Excitement! They have arrived at the entrance to the Moogh Zone.

On either side of the road are two armored personnel carriers bristling with machine guns. There are a dozen security personnel and Moogh Customs officers. Between the taxi cab and this formidable array of firepower, there are a dozen coaches filled with excited Flybys and three tankers full of Jet A1 fuel. As the cab approaches, the driver starts mumbling nervously.

"It's okay," Maggie tells him. "Go to the front of the queue." She withdraws her Moogh Zone ID badge.

They drive to the gate and Maggie lowers the window. She offers the card to the customs officer and he zaps it with a scanner.

"Welcome back to the Moogh Zone, Miss Tarp," the customs man says. Then he waves the taxi through the blockade.

The Genesis Amble

In the archetypal monster story, when the creature emerges from the forest, the townsfolk run away screaming in terror. The beast then sets about killing and wrecking havoc while the townsfolk organize, arm themselves, and set out for battle.

Exactly the opposite happens with the Moogh. Particularly for women, the typical reaction upon first seeing the Moogh is to run 'towards' it yelling for joy. In the early days, before there was effective crowd control, thousands of people flocked to the Moogh, swept up in mass hysteria characterized by screaming and swooning. They referred to this phenomenon Mooghmania.

Prior to the first sighting, no one had heard of the Moogh nor even conceived of the possibility of its existence. And yet, in the morning of January 3, 2009, the Moogh ambled out of a Greek cork forest and the first human to see it was a farmer riding a tractor.

Maggie Tarp was on a European holiday at the time, having just completed her journalism degree. Her bus was pulled to a halt when the driver saw the Moogh ambling across the road. It was being followed by a farmer, excitedly shouting and pointing from the seat of his tractor.

Maggie saw an opportunity, and she took it. While all the rest of the people on the bus were screaming with joy, Maggie interviewed the farmer. He described a feeling of intense empathy for the Moogh, as if it were the antidote to all the pain and suffering that he had ever experienced.

Quickly, word spread about this extraordinary beast that seemed content just to amble across the countryside, periodically stopping to hold its body in different positions that Maggie named poses. Within a week, the Moogh had a following of thousands. Most of the crowd were women and at first there was pandemonium, hysteria, fainting, crying and lots of fights breaking out as people struggled to get close to the Moogh. Some people even climbed on its body.

However, when a young woman fell from the Moogh and broke her neck, some of the followers organized and brought discipline around the Moogh. Those who followed the discipline were called 'Adherents' because they adhered to two simple rules: leave the Moogh alone, and protect it from interference.

The emergence of the Adherents was transformational. Quickly, the Moogh phenomena started to take on the appearance of a well-ordered cult. Hundreds of thousands of people flocked to the Moogh and were granted the lagom amount of access. Lagom means not too little and not too much; just the right amount. People were not allowed to get too close to the Moogh, lest the Amble and Pose of Moogh Destiny be interrupted. Nor were they kept too distant, for fear that they would wreck havoc.

The Moogh phenomenon was also hugely profitable for the few who saw the angles. Moogh Entrepreneurs were quick to realize, for example, that after the Adherents had spent the night standing in the rain watching over the Moogh, they would need a good meal and a hot shower.

All was going well with until one day the Moogh nonchalantly ambled across an invisible political boundary that separated Greece from Albania. The Moogh's passage was not, in itself, a problem for the Albanian Government welcomed the commercial opportunities that accompanied the Moogh.

The trouble started when tens of thousands of Adherents tried to amble across the border without visas. Very quickly the customs and police forces of two countries got involved, and the result was chaos and endless delays.

Then the scuffles began, and the riot police arrived by the truckload. The riots were intense, with dozens killed. The media showed iconic images of streets filled with smoke and people sitting on the sidewalks, clutching their bloodied heads after being bashed with batons. Others just cried pitifully for having been denied access to the Moogh. The incident reached a peak when the two countries mobilized military forces on the border and shots. The United Nations got involved when shots were fired.

The dispute garnered the attention of the world's media. It was the TV cameras and press photographers capturing dramatic images of pitched battles being fought around the pacifist ape that cemented the Moogh's international fame. Around the world, hundreds of millions of people tuned-in, glued to their screens, fascinated by the frenzied activity that this noble, passive creature had inspired.

The Moogh phenomena had begun, and millions of people set out on a Pilgramoogh, a journey to the Moogh for enlightenment, or Enlightenmoogh as it was known.

Just as the Moogh could be an agent of conflict, it could also be an ambassador for peace. This insight led to the establishment of a new United Nations department, a hybrid between the United Nations Educational, Scientific, and Cultural Organization (UNESCO) and the United Nations Environment Program (UNEP).

The first action of the newly formed United Nations Moogh Affairs (UNMA) was to pass the Declaration of the Adherents. This declaration created a Peace Park around the Moogh, a ten-kilometer radius zone in which internationally agreed rules applied.

As more and more people came to the Moogh, it became apparent that the UN bureaucracy was unable to manage the activities within the Peace Park. So it was decided to outsource the job. At the time, UNMA was chaired by the Russians and the bidding process was described by the then Secretary General as 'transparently opaque'.

Perhaps unsurprisingly, the management rights of the Moogh and the Peace Park were awarded to a newly formed not-for-profit organization called moogh.org, headed by Russian national, Maxine Slivkin. Overnight, changes were implemented, and the Peace Park got a lot less peaceful.

The militarization of the Moogh Zone was one of the defining features of moogh.org's management of the 11-foot tall peace primate and its followers. While the relationships between the people around the Moogh underwent a distinct transformation when moogh.org took over, the Moogh itself was unchanged in any of its behaviors of attributes. It continued to amble and pose as if it had not a care in the world.

The Long Future Diagram

After her long journey, Maggie finally reaches her hotel. She checks in and sits on the side of the bed searching the Moogh App for details of Maxine's presentation. It is a technical briefing on Long Future sustainability in a function room on a lower floor.

Maggie enters the auditorium where Maxine is running a powerpoint presentation. Maxine Slivkin, Chair of moogh.org, is tall and elegant and speaks with a Russian accent. She projects an aura of power and control that is at once both reassuring if you are on good terms with her and scary as all hell if not.

The presentation is already underway, and Maggie moves quietly to a vacant seat in the back row. She retrieves her special notebook and pencil from her bag and looks around the room to see who is attending. There are about fifty people, and she recognizes some as working with moogh.org. Others are journalists and bureaucrats and always there is the military presence. The military people are hard to ignore, physically large men in crisp uniforms. One of the soldiers Maggie recognizes as the Chief of Staff to the Big General. She makes a note of this in her book as he had once offered her an interview.

Maxine is addressing the audience using a green laser pointer to describe the elements of a diagram projected on the white screen. The diagram has both straight and curvy lines colored black, red and blue.

In the middle of the diagram is planet Earth, looking lonely and vulnerable with red and green dashed lines penetrating it like arrows. Where the arrows meet, there is a bright yellow flash, like an explosion. Swooping above and below the planet is a wavy blue line that continues to the edge of the screen where it terminates with an arrow head suggesting that whatever the blue indicates continues indefinitely. The diagram is at once simple and compelling.

Maggie leans forwards and listens intently, trying to catch up.

Maxine announces Maggie's arrival by telling the audience, "This is Maggie Tarp. Maggie works with Fractious News. She writes honest and empathic stories about the Moogh. She is a valuable part of this narrative. Welcome back, Maggie."

Maggie blushes as fifty people turn to look in her direction. She forces a smile and raises her hand to say hi.

Maxine continues, "Maggie, I was just telling our audience that this is a conceptual model called Long Future that puts a new perspective on the dialog around sustainability."

Maggie looks back to the diagram, feeling content that its mysteries are going to be revealed. The lines in the diagram remind her of positions of the hands of a twelve-hour clock.

Maxine continues, pointing to a solid black line running downwards towards 4.30 from the top left of the screen. She says, "This line represents natural capital by which we mean flora and fauna, forests, fish stocks, farmland, fresh air and fresh water. All forms of natural capital are in decline because of unsustainable human activities. The reason that human activities are unsustainable is complex, but the main cause is our outdated economic model called global capitalism."

Maxine points to the solid black line emanating from the lower left of the screen and running upwards towards 1.30. She says, "This line, going up, represents the increasing human demand for natural capital. It is going up because of a growing human population – that's seventy million new mouths to feed every year – and a global economy that doubles in size every thirty years or so."

Maggie observes that the dark black lines terminate in a small circle and that there is a gray arrow pointing to these circles with the word 'today'. There is also the words 'past' with an arrow pointing to the left and 'future' pointing to the right.

From the small circles, moving into the center of the big yellow flash are red and green dashed lines, representing a projection into the future. The red dashed line is an extension of the black line, and the green dashed line varies by a few degrees.

Maxine continues, "The red dashed line represents business as usual, a continuation of what we have been doing up till now. You will see that this leads to a crash in the global system. We refer to this as the Short Future, but it goes by other names including Collapse or the Great Disruption or Near Term Extinction."

One of the men in the room raises his hand and says, "This is new to me, what does that mean, Crash and Short Future?"

"It means a collapse in the global ecological system and the end of human civilization," says Maxine plainly. However, she sees that she has failed to communicate effectively to the man who asked the question.

Maxine thinks it through for a few moments before she speaks. "The living skin on this planet is a complex interplay of biological and physical processes that can endure for a very long time if we do not exceed critical planetary boundaries. If we exceed any one of these boundaries, the Earth will become uninhabitable for humans."

The man who asked the question is not satisfied, and he protests, "I understood about half of that. You know, you intelligent people need to learn how to talk to people who don't have rocket science degrees."

"Thank you and I apologize," says Maxine, "have you ever had a fish tank?"

The man laughs and says, "Yeah, I've had some fish tanks."

"Have you ever had one crash?"

"Crash?"

"One day it is a healthy living system with lots of diversity and color. And the next day, something goes wrong, and the tank turns into a glass chamber filled with dead fish and stinky brown water."

"Yeah, that happened once," says the man.

"What happened?" asks Maxine.

"It was really hot; there was a power failure, and the air pumps failed."

"That's a good example. Warm water holds less oxygen than cool water. There was no air pump to make up the difference and the fish asphyxiated. In this case, the amount of oxygen in the water is a boundary in the fish tank. You exceeded the oxygen boundary, by going too low, and this is what killed the fish. Are you okay with that?"

"Yeah, I get that," the man says. "If you exceed the boundary, the system dies."

"Exactly. There are nine planetary boundaries that allow human civilization to persist. And if we exceed the planetary boundaries, Earth's living skin could be reduced to a smelly, brown goo. And you have to question what your share portfolio will be worth if that happens. So it's not just an ecological collapse but an economic collapse that we face if we continue with business as usual. That's the red lines on the diagram."

"Oh, no," gasps Maggie, putting her hand to her mouth.

"Are you okay?" Maxine asks.

"This is a challenging subject."

"Sustainability is hard, Maggie."

"I understand."

"Losing a loved one is hard," Maxine says, "And losing seven billion people over the next few decades is also hard, but in a different way. That's why it is necessary to avoid the Short Future."

Maxine turns her attention to a man with his hand held

up. "Yes, John."

"You said that there are nine planetary boundaries. What are they?"

"That's the subject of the next workshop with Professor Fred Frotter. I'll let him explain."

"Fred Frotter, awesome," says John.

Maxine turns the laser pointer back to the diagram. "Okay, the dashed green line represents 'going green.' You'll see that making things marginally less bad just pushes out the date of the crash by a few years. Nothing particularly exciting about that from a 'save the planet' point of view."

Maxine observes the people in the room and sees them nodding their heads. Her voice changes to a lighter tone.

"Fortunately, there is good news. An option is to go Blue. Blue is the idea of a global economic system in which there are profitable enterprises that regenerate natural capital and reduce the human demands on that natural capital. You will see that the blue line, extending into the future eventually goes wavy, representing a natural balance. Regaining this natural balance will allow us humans to survive the inevitable shocks that come from natural processes."

"What sort of shocks?" asks the man in the front.

"Asteroid impact is a threat. Global pandemics are another. Large-scale seismic and climatic events like volcanoes, tsunamis, and super-storms. These things could prevent us from achieving a Long Future. Yes, John?"

"So where does the Moogh come in?"

"Excellent question, John. We think that the Moogh is like a canary in the coal mine. It's a wake-up call telling us that it is time to change direction. We have evidence that the Moogh Tears are highlighting that fact that we are crossing the planetary boundaries. I am not going to go into detail about this but if you come to the next workshop, you'll hear the full details."

Maxine listens to someone's question and then replies, "Frotter," she says, "Professor Frederick Frotter is running the next workshop."

"Awesome," thinks Maggie, excitedly. Frotter is a hoot.

The presentation ends and the people in the audience start to chatter as they find their way out of the room. Maxine comes over and takes the seat next to Maggie.

"Its good to have you back," she says, taking Maggie's hand in hers. Maggie observes the platinum and diamonds in Maxine's rings and necklace. The jewelry is exquisite and very expensive looking. "How was London?" Maxine asks.

"It was a long way from the Moogh." Maggie wonders whether she should disclose that the situation with Perrin and that he is probably sitting in a detention cell at the airport right now. She decides not to break that news just yet.

Maxine rolls her palm over Maggie's hand and says, "Tomorrow afternoon, I will accompany some scientists to see the Moogh. Shall we go in together?"

"Oh, yes," says Maggie excitedly. "Absolutely, yes."

Maggie's Routine

The next morning is the first full day in the Moogh Zone, and Maggie quickly gets back into her normal routine. The first task of the day is to check that the Moogh has made it safely through the night, and she does this sitting up in bed and checking the Moogh App. She finds that it has been ambling and posing overnight as normal. The security levels are all set to green, so all is good there.

Next, Maggie scans the internet for her favorite blogs, looking for any new ideas about the scientific, psychological, spiritual or environmental aspects of the Moogh. She avoids any links to mainstream media and particularly anything that is trending on the blogosphere. This ritual takes a minimum of an hour. Then she showers and takes a long breakfast in the hotel restaurant where she jots down her thoughts on various non-trending story ideas that few people will ever read.

After breakfast, she does the rounds of various individuals and agencies in the Moogh Zone. She swoops through Moogh HQ where she receives briefings from the various department heads.

Then she wanders to the Tent City where the Adherents live. This part of her day is always fraught with emotions as she remembers her time living as an Adherent and the circumstances that had helped her escape that grueling life.

Inside Tent City, Maggie flits about, seeking unstructured, butterfly-like conversations with those of the Adherents that she likes. She carefully keeps a distance from the more obsequious amongst them. Many times she has found herself caught up in endless conversations about how she had managed to break out of the cycle of poverty and drudgery that accompanies the life of an Adherent.

In the early days, the Adherents enjoyed spiritual freedom, fun and frivolous play. However, as time passed, and moogh.org took over and imposed increasingly rigorous discipline within the Moogh Zone, the party atmosphere waned.

Today, the life of the Adherent was not dissimilar to a monastic order with endless menial work and minimal reward, most of which remunerated with low-quality food and accommodation.

The frequent complaint amongst the Adherents was that they spent most of their time cleaning up after the day-trippers and little time in the presence of the Moogh. It was these rumblings of dissent that bolstered the ranks of the Moogh Underground, most of whom were disgruntled ex-Adherents, or Nonsticks, as they were known, colloquially.

These concerns were far from Maggie's mind today, however, because her relationship with Maxine meant that she could visit the Moogh just whenever she liked. Plus, as the senior Moogh reporter with Fractious News Network, she was well remunerated and lived all expenses paid in which-ever luxury hotel was closest to the Moogh Zone. Not a bad ticket, really, for someone with such a lackadaisical, non-trending existence.

On most days, she would find the time to take a trip inside the fence and get present with the Moogh. Sometimes she would be there on her own. In these instances, she'd fall into a tranquil, meditative state and just quietly observe the Moogh ambling or posing.

However, on most of her visits inside the fence she accompanied various groups that visited the Moogh. There were three main categories of visitors inside the fence.

These included the scientists, of which there were many and varied. They seemed to be almost continuously poking, prodding and sampling various bits of the Moogh. While the Moogh didn't seem to mind, the constant intrusion was a constant source of frustration to Maggie.

There were the groups of Adherents, getting their

monthly half-hour presence with the Moogh. They would be herded inside the fence in groups of fifty, surrounded by a phalanx of Level-4 security. Often, there would be bouts of Mooghmania, lots of screaming and passing out. Sometimes, the Adherents got out of control and the Level-4 zapped them with their stun guns. Other times, the Adherents were more composed.

The third class of visitor to get face-time with the Moogh were the special folk of one type or another from moogh.org or UNMA. Typically, Maxine would escort these people to the Moogh and there would be champagne and canapés before and after the visit.

On a typical afternoon, after meeting the Adherents or visiting the Moogh, Maggie would find her way to a cozy restaurant for a write-up. She would spend hours drinking herbal teas and writing stories in her special notepad that she would eventually type up and submit to the Fractious News Network Moogh Desk. Typically, she would write two, 500-word stories a day; sometimes more, sometimes less. These stories would form the core of the Fractious News Moogh coverage, with the rest being general Moogh stories from over the wire.

Typically, after submitting her non-trending stories, Maggie would spend an hour or so with the moogh.org executives, sharing her findings of the day.

Today, her afternoon involves a date with Maxine and the Moogh. To this end, Maggie finds her way to the lobby of the Plush Hotel at three pm.

The Moogh Looks Like

The Moogh is the size of a bear walking on its hind legs although it clearly isn't a bear. Golden colored fur cover its back and sides, and its face, chest, belly, and groin appear to be naked, just a beige colored skin. However, on closer observation, the bare skin is super-fine hair that seems to shine, not unlike that on the pelt of a sea lion, even though it is not wet.

The Moogh lacks external genitalia with just a smooth rounded bump where the male or female parts would otherwise be. There is much speculation about what that bump might contain. Some say male bits, others say female bits, it all just depends on whether you think that the Moogh is a male or female. The truth of the matter is that the Moogh is neither fully male nor fully female, seeming to have transcended the troublesome issue of gender altogether.

Women typically identify the Moogh as male, mainly because of its apparent strength and size. It has lithe, muscular limbs and strong cheekbones. Despite these masculine features, it is ascribed a distinctive type of maleness, described in three words: 'safe for women'. Indeed, the Moogh is safe for everyone, except for those foolhardy enough to climb on its back then fall off onto the hard ground. It does not move fast, it has no appetite, and apart from creating hoards of avid followers, all it does is amble and periodically throw a pose.

It is not so easy to describe what the Moogh looks like because it is a completely distinctive species, and it has characteristics of a variety of different animals. One could say that it looked like a bear because of its size and bipedal posture – but it looks nothing like a bear. You could say it looks like a cow, because of the sad, doleful eyes that draw people in – but the Moogh looks nothing like a cow. Some have even likened it to a sea lion because of its skin-like fur. Everyone who sees the Moogh creates their own story about what it looks like. However, the one thing that everyone agrees on is that the Moogh looks like the Moogh.

The most important thing about the Moogh is not what it looks like, but how it makes people feel and behave. With the exception of the screamers who start screaming and don't stop until they pass out, for most people who see the Moogh up-close, the response is the same.

In the first instance, they gasp in awe at the uniqueness of the creature as they try to make sense of something that they have never seen before. A cathartic sigh - '*awwwh*' - quickly follows this first gasp. Seeing the Moogh gives one an immediate sense that everything will be all right, after-all.

The way that people behave around the Moogh is unique, and the closest analogy is the response that Western people have when they come across a whale or a dolphin stranded on a beach. They flock together in groups and do their best to help, irrespective of how difficult the conditions are. Common people become heroes in these situations. They discover an innate reserve of strength, goodwill, and cooperation. The Moogh engenders this same response, a sense of awe, reverence, and deep respect.

The other thing to notice about the Moogh is that it is never alone. Not only humans flock to it, but so do other species. It is forever surrounded by living things with wings, birds and insects of hundreds of different kinds. They flit and swoop around the Moogh like they are celebrating its very existence, or forming a protective aerial service like a squadron of Bahama swallows around a critically endangered Californian condor. The noise they make is distinctive and audible from a long distance away. It is a symphony of chirps, squawks, squeaks and every other noise ever emitted by a bird and the buzz of fast-flapping wings.

It is from this incredible airborne opera that the Moogh got its name. In memory of the famous musical synthesizer, the Moog, that was able to concoct so many different types of noise.

The name quickly spread through the original Adherents, and it was a sticky and very soon no one questioned that the Moog was the name of this extraordinary beast. The variation came later when the 'h' was added (by Maggie Tarp – then known by her pen-name Marequa Taard) to distinguish it from the musical instrument and to put her indelible stamp on the phenomena.

Of all the Moogh's flying companions - collectively referred to as the Entourage - it is the bees that get the most attention. The bees feed on the Moogh Daisies that grow in profusion where ever the Moogh goes. It is as though the very touch of Moogh feet on soil turns the ground into a fecund flowerbed impatient to sprout; the Daisies grow within minutes of the Moogh's passing. And when the Moogh stops to pose – if it poses for too long – it is in danger of becoming lost in the yellow and orange flowering foliage that grows around it.

As a result of this profusion of nectar-rich flowers, the bees are distinctly large and healthy and numerous around the Moogh. As a result, the Moogh Zone is dotted with bee hives, some natural, some man-made. Inside these hives is grown a magnificent honey, Moogh Honey, which is rich and golden and has all the attributes of good honey and more.

Moogh Honey is used to make a delicious and intoxicating brew called Moogh Mead. Those who drink Moogh Mead at night report dreams of a utopian society. In this world, there are no fossil fuels heating up the planet, dead zones swirling in the seas, or plastics choking the world beaches.

In these Moogh Mead dreams, the living systems of

Earth are not being crushed by a rapacious economic system headed by a global superclass elite who are so unplugged that they are not just out-of-sync touch with nature, but ignorant of their responsibilities as humans, as well.

Instead, in this dream world, the biosphere is regenerated, renewed, regrown by a nurturing economy. The human population lives rich and full lives in balance with nature. This is an achievable world, where corporations take full responsibility for their products – across the entire life cycle. Sustainable plastics are grown in bioreactors fed by nutrients from cities. And when these bioplastics find their way into the ocean – as plastics of any type inevitably do – they become a meal for marine life rather than a death sentence.

The hangovers from Moogh Mead are relatively mild, as well.

The Moogh, at Last

Maggie meets Maxine in the hotel lobby, and they step into an electric Mercedes limousine that pulls up in the foyer at the allotted time. The cabin of the limo is plush with white leather seats illuminated by computer monitors that report on the Moogh and all the activities that take place around it.

The limo is a mobile office for the moogh.org executives and also play room. To that end, Maxine mixes two vodka orange drinks from the minibar and hands one to Maggie. Resting back in the comfy seats, she smiles warmly and pushes her fist into the soft flesh on Maggie's shoulder. "Its good to have you back, kiddo."

"Its good to be back. What do the scientists want with the baby today?"

"They are going to snip some fur."

"What for?"

"They are looking for heavy metals."

"Really?"

"They think it will shed light on its geographical origins. There is a security briefing first, but I expect that you'd rather feel the crowd."

"I'd rather feel the crowd," Maggie says. She sniffs the vodka, unsure of the consequences of drinking afternoon alcohol. She takes a sip and rests back, contemplating the odd look that Maxine has on her face, as the tall Russian looks her over.

Fortunately, the trip to the Moogh Fence is over quickly. The electric limo pulls to a halt and Maggie steps out, feeling light-headed and euphoric from the vodka. She moves towards the chest-high, white picket fence that surrounds the Moogh.

There is a crowd around the fence that is three deep and constantly jostling for position. Maggie walks around the perimeter observing the crowd until she finds a place where she can lean against the fence and see the Moogh.

The Moogh is a kilometer away, a little more than a dot in the distance. It stares skyward, completely still, almost like it were a statue of a stage actor in the middle of a speech, gesticulating towards the clouds. Maggie retrieves her smartphone and checks the Moogh App. She sees that it is holding Pose #4 Stargazer.

"What's it looking at?" asks a man standing next to Maggie. He is holding his young daughter, a little blonde-haired girl with a runny nose who sucks her thumb peacefully. The child is holding a Mooghie, one of the official Moogh Toys available through the gift shop. Mooghies are notoriously expensive and heavily promoted for the nine different noises they make when one presses the buttons hidden under the fabric.

Maggie knows the locations of each of these buttons including how to activate the ninth noise that most kids never find. She raises her hand and squeezes the Mooghie in the special place, and the speaker buried in the doll's chest lets out the infrequently heard 'ninth'.

The Ninth is difficult to describe, but it sounds something like a cross between a kitten meowing and a child laughing. It had always seemed strange to Maggie that the Moogh, an animal that had never once uttered a single sound, should be represented by a toy that was so overtly vocal.

The child breaks into an infectious giggle and hugs the Mooghie closer to her body.

"Huh! I've never heard that noise before," the father says, surprised.

"We call it Moozart's Ninth," Maggie tells him, chuckling at the joke. "And the Moogh is just gazing," she continues, answering the original question.

"But what's it gazing at? There's nothing up there."

"Maybe there's nothing up there to us," says Maggie. "The Moogh has its own ideas about what is important."

"You mean there's something we can't see?" the man asks.

"Maybe we can see it, but we don't interpret it as important."

"I guess," the man says. "Thanks."

Maggie continues her journey around the perimeter fence and eventually comes to the security gate where Maxine watches three scientists getting patted down by Level-4 Moogh Security. Maggie allows the security guard to scan her ID card, confirming her on the list of people approved to get close to the Moogh.

Eventually, all five are cleared, and the gate opens. They follow Moogh Approach Protocol and walk to a point where they are directly in the path of the Moogh. They move slowly in a single-file directly towards the Moogh. Maxine takes the lead, followed by Maggie and then the three scientists, dressed in white coveralls and yellow, rubber boots. When they are five meters from the Moogh, Maxine and Maggie halt and the scientists approach the Moogh and collect samples.

The Moogh stands about four meters in height. It is often described with reference to a standing gorilla even though the differences are many. The Moogh is not as masculine as a gorilla. Quite the opposite, it has a childlike, almost feminine aura. Its skin looks like soft, beige leather with a fine hair all over. When the light is low, particularly in the evenings, the Moogh has a golden glow. The Moogh is not obese, but it is a bit portly. It has the physique of a bookish Singaporean kid who has eaten too much fried chicken. It is a little out of shape but otherwise healthy. On its belly, there are three horizontal lines of darker hair and from a distance, these give the Moogh the appearance of a six-pack. No wonder the ladies take a fancy to the Moogh.

Also, unlike a gorilla, the Moogh's face is very gentle. It has a humorous expression, like a newborn child wearing a look of pleasant surprise. The hair on the Moogh's face is longer and darker than on the rest of its body. It has shaggy eyebrows and tufts around its small ears. The Moogh's big, drowsy eyes cause many people to form an immediate affection for it. People want to coddle the Moogh and make noises near it to see if it will raise its eyes towards them.

However, never once has the Moogh purposefully looked at a human being. Many times, particularly in the early days when access to the Moogh was unregulated, people tried to attract its attention by standing in front of it and waving or calling out. But in none of these instances did the Moogh ever grace that person with a glance in their direction. It is as if humans are invisible to the Moogh; as if humans did not exist. This behavior of the Moogh has not gone unnoticed amongst the Adherents, and it is the one point that sometimes makes them turn away or speak against the Moogh.

Even though the Adherents are empaths, and they follow the Moogh for the Moogh's sake, they are also human, and everyone needs recognition sometimes. Early on, someone commented that cows had this same habit of ignoring people and from that point, the word cow was used pejoratively to describe the Moogh, even though the Moogh looked nothing like a cow.

As the Moogh continues its slow amble, Maggie and Maxine watch in reverence. As it ambles, it leaves behind it a trail of freshly grown Moogh Daisies. They spring from the ground in full flower, and are quickly taken over by plump bumble bees.

One of the scientists swoops up some of the bees in a net and drops them into a killing jar where he adds a squirt of CO_2 gas. After a few seconds, the bees are dead, and he retrieves one and shows it to Maggie and Maxine.

"Look at this one," he says, holding the bee by one of its legs. "Look how fat it is."

"He is a plumper, alright," says Maggie, intrigued.

The scientist nods his head as he inspects the fat bee. He tells her, "Before the Moogh arrived, the bees were dying out. This threatened to wipe out hundreds of types of plants including many that humans need for food because the bees are avid pollinators. But now the Moogh is here, the bees are safe and the forests are growing back."

The scientists go about their work, delicately cutting little pieces of fur and scraping a special tool against its skin to collect samples. The Moogh ambles along, showing no sign that it even knows they are there.

"Do you think that it will ever look at a human?" Maggie asks Maxine.

"One day," says Maxine, "I am sure of it."

After an hour of being present with the Moogh, the women return to the Plush Hotel in the electric limousine. In her room, Maggie sets up for a writing session.

She retrieves an immaculate wire bound notepad made of hemp and organic cotton paper and embossed with a Moogh Emoticon on the cover. Tucked into the recycled and recyclable metal binding is a very sustainable pencil made of apple-wood and charcoal. She scribbles a spiral in a circle at the top of the page to gauge the sharpness of the pencil. She has the muse, and she has the tools. Now she is ready to craft some words.

Maggie writes fleeting, wispy strokes of the pencil, leaving a tangle of shapes and curls in exquisite shorthand. She will type the final version of the story on her laptop then email it to the Fractious News Network editors. But this is no reason for the first draft to be anything other than immaculate, hand-crafted and beautiful. Her story brings in the little girl with the yellow hair and the observation that the Moogh senses things that the humans don't.

Moogh Tears

While Maggie is writing her story that is destined not to trend in the blogosphere, the Moogh halts its amble and adopts the #2L pose, colloquially referred to as 'my left shoelace is undone'. It leans forward, looking towards the toe of its left foot with its left hand held slightly forward.

On the ground, in front the Moogh's foot, is a dead bird, probably killed by a speeding car. Its neck is broken, and its tiny tongue protrudes from its mouth. It has yellow feathers with flecks of aquamarine and gold around the eyes. The feathers are ruffled, but the bird is otherwise intact. Its little eyes are open and lifeless.

The Moogh is motionless, its eyes trained on the bird. In the distance, behind the perimeter, thousands of people hold their breath in anticipation. There is a rumor circulating. Amongst the people in the crowd, there is a growing agreement. And then someone says the words, "The Moogh is going to cry."

The Moogh starts to move from its motionless pose. It is a slow but distinct motion. The Moogh moves its head up and to the right and makes a noise as if it were breathing tears through its nasal passages.

The people in Moogh headquarters who manage the Moogh App huddle around a monitor showing a video feed from the field. They observe the Moogh carefully and then call it: the Moogh is crying.

They activate the script that sends out a push notification to all the Moogh Apps around the world. Simultaneously, three hundred million smartphones ring out a simple tune, the sound of a small brass bell being struck four times: Ting. Ting. Ting. Tinggggg.

Across the Mooghsphere, hundreds of millions of people peer into their smartphones and begin to weep in sympathy with the Moogh. Some take the opportunity to find a quiet place and cry alone. Others turn to their friends and cry together, hugging each other as they sob.

The steely Moogh Drone operators aren't crying, however. They are busy lifting the drone off its platform and directing it through the sky, beaming live video feed of the Moogh, with a telephoto lens focused on its face. Its tears become visible as they express from the rim of its eyes and absorb in its fur.

The weeping is most intense amongst the Adherents who are by now well versed in mass-crying. They flock together in great weeping-circles and bawl their eyes out. The anguished noise of all that slurping and interrupted breathing, combined with the sight of all the snot, red faces and tears feeds upon itself and becomes deeper and harder.

In her room, Maggie hears the chimes from her phone and opens the Moogh App. Its effect on her is instantaneous. She feels a reddening of the face; heat builds up around her eyes and her tears start to roll. When she cries, she thinks about every bad thing that that has ever happened, and she sobs it out.

Across the planet, the pandemic of tears spreads. Hundreds of millions of people are set free from cultural taboos of showing sorrow in public without any particular reason. From New York to Sydney, from little villages in South America to the great sprawling metropolises of East Asia, the Moogh has made it okay to cry in public without people thinking of you as weak.

The Moogh crying is cathartic. The tens of millions who participate in the practice feel life's worries fall away. Upon conclusion of their ritual, they get straight back to work or whatever it was they were doing, feeling lighter of spirit and relaxed.

While the Moogh can bawl for hours, the humans are more easily satisfied. Three minutes is the average time for Americans. The Brits a little longer, but they don't do it with such vigor.

Maggie's weeping is subtle and personal, and it lasts on

average two and a half minutes. She mops her eyes and sighs a distinctive "Hoo!" as the concluding part of her ritual. Her regular after-weep routine of lounging around and feeling relaxed is interrupted by an incoming text message. She checks her phone and sees that work beckons. It is a request from Moogh Radio for her to attend an interview.

A Gestalt of Gaia

The Moogh Radio studio is located in the back of a truck that moves wherever the Moogh goes. It is a mini nerve center with satellite dishes and data streaming in from all over the Mooghsphere. Today, the journalist is an old hack from the BBC. He is gaunt with pale smoker's skin. The inside of the truck smells like an ashtray and Maggie grimaces as she steps inside.

She thinks that she could cancel the interview, but then that would most likely get back to the Dim Director. Moogh Radio is good promotion for Fractious News.

The hack announces her arrival, "We have with us today, Maggie Tarp from Fractious News Network. How are you today, Maggie?"

Maggie slips the headphones over her head, frustrated that the radioman is hurrying her along. Usually, she'd take a few seconds to adjust them properly so that they didn't pull her hair.

"I am feeling Mooghy, today," she says, honestly as the cry had done her a world of good. She grimaces, noticing that she has rested her arms on a surface and now has cigarette ash on her sleeves.

"Today is the sixty-third Moogh crying event. They happen about once a month, on average. Is this the first time the Moogh has sniveled over road kill?" the journalist asks, abruptly.

Maggie is unimpressed that the man has referred to the Moogh as having 'sniveled over road kill'. It is a crass expression that trivializes the profoundness of the event, and he is apparently trying to get a rise from her. She wants to take him to task but decides to let it slide. So instead, she makes a little smile demonstrating that she is not taking his bait. Then she replies using perfect Moogh English, "The first time that Moogh Tears were drawn by something killed on the road, it was an otter."

"A flat otter?"

"A very rare otter, poor thing, threatened with extinction. But this is the first time that a dead bird has drawn Moogh Tears on the road."

The radio man looks Maggie over with that same hungry expression that the Dim Director uses, and Maggie has no option but to suffer the indignity in silence. She keeps the grin affixed to hide her real emotion, which is that of pity for the wretched man, stuck in this box full of electronic machines and his dirty habits.

"I keep hearing that the Moogh Tears are an empathic response, tell us what that means," the old hack says.

"It is believed that the Moogh is picking up on the collective mental energy of the Empaths amongst us and allowing these sentiments to be understood by other people."

"And what is an Empath?"

"An Empath is someone who has empathy for people and things other than themselves." Maggie looks around the radio desk trying to locate the source of the ashtray, but it seems to be hidden by documents spread messily around.

"So, aren't all humans Empaths?"

"To a degree, yes. But most people care most distinctly about things that have a direct influence on their lives. They care about the families, their workmates, and possessions. They have no real empathy for abstracts and concepts like biodiversity or the ozone layer, for example."

"So the Empaths are people who care about the Moogh?"

"Empaths are people who care about everything including the Moogh,"

"That must be exhausting," say the radio man, snidely.

"Empathy has its own energy supply," says Maggie, "so you are always fully charged."

"So the Moogh is telling us the things that we should care about. Acting as a guide as to how we should live our lives on Earth?"

"That's the theory that has been trending for a while now," says Maggie. "And it's a Blue Ocean theory. No competing theory comes close to it for popularity amongst the Adherents."

The video monitor shows that the Moogh has started to wipe its eyes with the side of its clenched right fist. It doesn't wipe the tears away, so much as allow them to absorb into the fur on its hand.

The radio man asks Maggie, "Now the theory about this eye drying technique is not Blue Ocean, is it?"

"No, there's about a dozen ideas, and new ones all the time. Lots of competition here."

"Is there any theory that trends more consistently than the others?"

"One that keeps popping up is that the Moogh is transferring some of the lacrimal fluid to the contents of its hand."

"Lacrimal. That's a big word for what exactly?"

"Yes. Sorry," Maggie remembers the Dim Director telling her off for the same thing. "Tears. Lacrimal fluid is tears."

"So, it's transferring tears to the thing inside its hand."

"Assuming that there is something in its hand, yes."

"And no one knows why."

"That's right."

"So whatever it is, it needs to be kept moist? Is that how we read this?"

"That's one of the theories," says Maggie.

"Fascinating," says the radio man, unconvincingly.

The video feed in the radio truck shows the Moogh make one long sniff and then the crying stops. It raises its head and then starts to amble again, moving slowly down the road.

Around the perimeter of the computer monitor, there are strings of data. The radio man looks at the numbers and says, "Okay they have called it. That's the end of the sixty-third Moogh crying event. It lasted for one hour twenty-three minutes and twelve seconds. And the Moogh was crying over a dead bird on the side of the road. This is Moogh Radio, and we have Maggie Tarp with us. Maggie, from where did the Moogh come from?"

"That's the million-dollar question."

"One minute there is no such thing as a Moogh and next minute 'poof' it just shows up, wanders out of a Greek cork forest. Its genes don't match anything known to science. It has an influence on people that has never been observed before – nearly a billion Facebook friends – and people flock to it like whale watchers on steroids. Where's it come from, Maggie?"

"There's lots of theories. It came from under the ground as a result of the global warming. It came from space. There is a theory based on quantum physics that it popped into existence as a consequence of an interference pattern of subatomic particles. There is even a new theory that there is, actually, no Moogh."

"No Moogh? That's a good one. Maybe the Moogh is God?" suggests the radio man, provocatively.

"Oh, no," says Maggie, instinctively. "Although there is some interesting research into the way that religious people relate to the Moogh.

"Can you share that with us?"

"This is not official. It is not from the leaders of these religious organizations, but random phone polls. The results show that the Buddhists believe the Moogh to be a manifestation of the Buddha. The Bahias think it a new heavenly messenger, although they don't know what the message is. The Christians regard it as another of God's creatures over which they automatically have dominion. Muslims think it's a demon. And the Jews decline to comment."

The old journalist starts to laugh and Maggie grimaces as

he shows off his yellow stained teeth. "And what's your view on this?"

"I'm with the atheists on this one. I think that the Moogh is from Gaia."

"This is Gaia Theory you are talking about?"

"Yes. Gaia theory says that all the living things on Earth are in-sync with one another, and they help create a self-balancing system. We humans are part of nature but through our cultures and technologies, we have fallen out-of-sync with nature. We've disturbed the balance. I think that the Moogh is a gestalt of Gaia that is showing us how to get back in-sync with nature."

"A gestalt? Is that like a synergy?"

"It is similar. A synergy is something 'greater' than the sum of its parts. A gestalt is something 'other' than the sum of its parts."

"And what's it doing here?"

"I can tell you what I think."

The report says, "This is Maggie Tarp, Senior Moogh Reporter with the Fractious News Network on Moogh Radio. So Maggie, what is the Moogh doing here?"

Maggie feels like she is on the home stretch, just seconds away from departing the confined, odorous space. She chooses her words carefully, hoping not to provoke any more questions. She leans forward, close to the microphone and places her fingers on her headset.

Then she says very clearly, "I think that the Moogh is a physical manifestation of Gaia that has been formed to show humans how to live in balance with nature. That's why it is here."

A Prior Assumption

After the interview, Maggie returns to the Plush Hotel to check in with Maxine. Entering the suite on the upper floor that forms the offices of the moogh.org executives, she halts, sensing that something is out of place. Then she hears a sound that she'd hoped never to hear again.

Maggie cautiously peers into the lounge area to see that Perrin has found his way into the moogh.org executive suite. He is lounging in a comfy chair and has the rapt attention of Maxine and a few of the other senior figures in the group.

Perrin is waxing lyrical about how he failed to get through customs. "So I'm saying to the customs dude, I'm saying, dude, I'm a Moogh reporter. Doesn't that mean anything in this country? I don't need a goddam passport or a driving license or a SIM card in my phone."

Perrin sees Maggie peering at him, and he turns his full attention to her, raising his voice. "I don't need none of that stuff to avoid a stretch in jail and a severe skype bollocking from the Dim Director." Then he announces, with theater, "Hey! It's my assistant, Paprika Tart! Howdy cowgirl!"

"Grow a womb, Perrin!" Maggie snaps.

"So he is with you?" asks Maxine, surprised, turning to Maggie.

"I wish I could honestly say that he wasn't, but technically he is," Maggie replies, cryptically.

"That's a yes, I assume," says Maxine, confused. "Your associate tells me that he is to be granted a security pass."

"Director's orders," says Perrin, smugly.

"Which are what exactly?" Maggie asks, defensively.

"The Director says that Maggie's on the next flight home if she doesn't cooperate."

"We can give him a pass if that's what you want," says Maxine.

"It's not what I want to do," says Maggie. "It's what I have to do if I want to keep my position."

"I know I can come across as a bit brash," Perrin says with a faint hint of humility, "but I'm a good journalist."

"I'm sure you're a great journo," says Maxine. "I'm just not sure that your empathy will resonate with what we are trying to achieve here. What do you think, Maggie?"

"I am convinced that his empathy won't resonate," she says, plainly. "That ought to be an *a priori* assumption around Perrin Speer."

"So," continues Maxine, "the question is, which is the lesser of two evils? No Maggie or Maggie plus her annoying companion." A few long seconds pass while she thinks it through then makes her decision.

"So, Maggie, I will agree to a security pass for Mr. Speer, and I will leave it up to you to decide when it should get revoked."

Maggie is instantly relieved to have been entrusted with Perrin's fate. She can have him escorted out of the Moogh Zone by Level-4 Security at a moment's notice.

"And as for you, Mr. Speer," says Maxine, pointing one of her manicured nails in his direction. "Do not piss me off. I have an oxidizer inside."

"What does that even mean?" asks Perrin, taken aback, as he watches Maxine depart the room with her companions, leaving him alone with Maggie. He starts waving his hand, "No, no, no, Paprika, I won't hear of you apologizing for screwing me over at the airport."

"Well, we agree on that at least."

"So, let's go and get that ID card, then, shall we? Where's that?"

"That's at Moogh HQ," says Maggie, glumly.

"Moogh HQ. After you."

Moogh HQ

Moogh HQ occupies a large military tent a few kilometers from the Moogh Fence. Moogh HQ is the official nerve center of the Mooghsphere. A United Nations task force, contributed to by nearly every country in the world, run it. The bustle inside the bivouac is a multi-cultural affair.

The central feature inside the tent is a command center that resembles a control room at NASA during a rocket launch. There are rows of desks with computers, machines and droves of technicians monitoring them. There are scientists and government officials, all of whom have a particular interest in studying or managing the activities around the Moogh.

Onto a wall, an overhead projector shines a huge Mooglemap - a google map with added layers of Moogh data. A color-coded line shows the speed and direction of the Moogh's amble, as well as statistical probabilities of where it will go next. Little icons on the path line show where the Moogh stopped ambling, and the different poses that it struck.

Around the perimeter of the Mooglemap are panels showing data including weather, and predictive assessments of where the Moogh will amble next allow. This data allows for the orderly relocation of Moogh Infrastructure, as well as the removal of fences and other obstacles to the Moogh's amble.

Relocating the Moogh Infrastructure is no small task because there is so much of it. It includes Tent City where the Adherents live, the security perimeter that extends for miles around the Moogh, Moogh HQ and all of the equipment therein.

Maggie pulls Perrin to a halt inside the door. "What the hell?" he says, taken aback. His mouth hangs open as he absorbs the scale, complexity and order of Moogh HQ.

Maggie moves over to a young woman who wears moogh.org cap with a ponytail sticking out the back. She carries an ipad resting on a blue folder. "Hey, Maggie, you're back," the woman says warmly.

Suddenly, there is the sound of a bell ringing and all the busy people in the bivouac respond by looking towards the Mooglemap. The icon that represents the Moogh is now moving, and the line that follows indicates a speed 26% higher than the 48-hour average.

"He's in a hurry today," says the lady with the ponytail. "That's going to make life really interesting."

"This is Perrin Speer." Maggie indicates towards her associate. "He is working with me and needs a security pass."

"Sure. He'll need to sit the Moogh induction first."

Maggie directs Perrin to a room where there is a dozen fold-up chairs and a data projector shining onto a white canvas.

"Do I have to?" Perrin protests, "I'm already an expert."

"If you want a security pass, you'll have to sit down and shut up."

Perrin grunts and slumps onto one of the metal chairs. He crosses one leg over another and strikes a pose that suitably illustrates his frustration. The induction movie flickers onto a white screen.

The video shows a team of scientists in biohazard suits moving very delicately around the Moogh. They take samples of hair, surface moisture and temperature readings. One has a camera with a frame around it. He pushes the frame onto the Moogh's flesh and then covers it with a black sheet. Under the sheet, the camera flashes, providing high resolution close up of the Moogh's skin.

"What's this?" Perrin asks, suddenly interested.

"This is when the Moogh first appeared. The World Health Organization ran tests to see if it is a threat to human health."

The movie continues, and there is a sequence showing the United Nations General Assembly debating the Moogh. Hundreds of delegates raise their hands as the Declaration of the Adherents is passed.

"And what is this all about?"

Maggie explains in a well rehearsed tone, "The Moogh is trans-boundary, it has no sense of political lines on a map. When it is in one country, that Government wants to control it. Then it ambles across an invisible political boundary, and there is a massive argument on the border. The Declaration afforded the Moogh international protection. It did this be establishing a Peace Park within a ten kilometre radius of the Moogh, inside which a set of internationally agreed rules applied."

"Did all the nations agree?" Perrin asks.

"All but one."

"Let me guess," says Perrin. "The good old U. S. of A."

"Yes," says Maggie. "The one nation that could change the trajectory of the global economy to make it sustainable is controlled by an elite of fossil-fuel addicts and religious people. So naturally they objected to something that promoted peace and respect for nature. But it was a vote in the General Assembly and the US didn't have a veto."

"So how does this work then, Mooghsphere, Moogh Park, Moogh Zone?" asks Perrin, "It's all just mumbo-jumbo."

"Only because you don't know it."

"Well help me out, Paprika, I'm dying here."

Maggie moves over to a cupboard and retrieves a small booklet. She hands it to Perrin.

"This is the dictionary of Moogh English," she says.

"It's got your name on the front."

"That's because I wrote it."

"I want a signed copy in hardback," Perrin says, cheekily, flipping through the booklet.

"You'll need to study that to report on the Moogh," she

replies, tersely. "And just to get you started, the Mooghsphere is everything to do with the Moogh. All the things, words, ideas about the Moogh. The Mooghsphere didn't exist six years ago, and now it spreads around the world through the media. The Moogh Zone is the Peace Park that extends across a ten-kilometre radius from the Moogh and where the international pacifist rules apply."

"And who's allowed near the Moogh?"

"Moog.org staff. The Adherents. They are allowed in on a roster basis. Scientists. Security people. Buddhists. Others at the discretion of moog.org." Maggie looks at the floor, frustrated by what she has to say next. "So, we'll need to get a NOTAM."

"A Notice to Airmen?" asks Perrin, confused, "What? Are we flying somewhere?

"Moogh English," Maggie growls, "look it up."

Perrin flicks through the dictionary and finds the entry. "Notice to Approach Moogh." He shakes his head wearily. "So, we're going to see the Moogh?"

"Yes," Maggie says, glumly.

Perrin heaves a sigh. "Great. So what happens next?"

"You stop talking and watch the rest of the video."

The Moogh with Perrin

On the way to the Moogh Zone, Maggie opens the Moogh App and fills in a NOTAM, requesting an approach with Perrin. The application goes straight to Maxine's smartphone, and the reply comes back quickly. It is a recorded video, and she says, "You can take him inside, but I want a pair of Level-4 security guards on him. If he makes one false move, he'll be stunned and banned from the zone indefinitely."

"Good work Perrin," says Maggie. "You've pinged a security detail."

"Good. They'll protect us from the thing."

Inside the security gate that leads inside the fence is the Level-4 Moogh Security. They are powerful, voluptuous females clad in the latest military fashion wear: white Milfash catsuits. Their red cork boots reach past their knees, and the similarly colored cork straps hold a variety of lethal and non-lethal weapon systems to their bodies. The most noticeable of these is the Rhinox-25 stun gun, so named because it can knock out an angry rhinoceros at twenty-five yards. They also carry a Walther .38 caliber pistol tucked into a pouch under the left breast. A 'misericord' carbon-steel trench knife is sheathed in a red cork scabbard connected to the waist strap.

"Holy hardon!" stammers Perrin when he catches sight of these deadly, combat beauties. He tries some ineffectual charm to ease his nerves, "Howdy?" he says, reverting to his cowboy analogy.

The Level-4 offers a steely glare and directs a tightly worded question to Maggie, "Will the male cause trouble?"

"Most likely," says Maggie, "I'd keep the stun gun trained on him if I were you."

"Thanks for that, Paprika," say Perrin. He holds a grim look on his face as the Level-4 pats him down for weapons. "It's not a gun, I promise," he quips, "I'm just pleased to see you."

"You're in my sacred place now," Maggie tells him, firmly, "act accordingly." She says this with a tone so terse that it surprises her. She thinks that she might be channeling Maxine's admonishment to 'not piss her off.' Then she's reminded of her conversation with Novell about this very instant, the point in which she takes something abhorrent to her sacred place. Novell's advice was to offer a sacrifice there. Maggie glances at the weaponry carried by the two guards. Maybe they'll do the job for her, she thinks.

"Follow me," Maggie finally announces, once her ruminations have come to an end and Perrin has been declared safe to enter the fence. "Make no sudden movements and keep quiet."

"That sounds like me all over," says Perrin, chuckling.

There are three hundred meters of open grassland between the security gate and the Moogh. Maggie walks ahead, Perrin follows, and the two Level-4 guards bring up the rear, their stun guns trained on his back.

The grass is, damp and the trudge seems to take forever. Eventually, Maggie halts five meters away from the Moogh. It is motionless, silent, holding a pose. She takes a step back towards Perrin and whispers, "We call this pose Swan Dive."

The Moogh is motionless even though above its head is a swarm of birds and insects buzzing, flapping and making distinctive noises. Around the Moogh's feet, the ground has erupted in Moogh Daisies, and these are picked over by hundreds of healthy looking bees.

Perrin looks the Moogh up and down at close quarters, showing an unimpressed scowl. He observes its clenched right fist. "What's it holding in its claw?"

"*Shhhh*," Maggie hisses at him.

"Okay," Perrin says in a quieter tone. "What's it holding?"

Maggie takes his arm and moves him back a few meters.
"You're not supposed to talk around the Moogh."

"It's a simple question."

"No one knows what it has in its hand," she whispers "Its held it like that from the very beginning."

Perrin goes back to silently observing the Moogh. He gives it about a minute and a half before he is compelled to speak again. "Is that it?"

"Is that what?" whispers Maggie.

"Is that all it does? I think its battery's gone flat."

Maggie feels her hackles rising. Why can't he just be reverential like everybody else? Why can't he just revel in the presence of the Ninth Wonder of the World? Why can't he just *Ughhh!*

"What else do you want it to do?" Maggie snaps.

"Well, it could move, for example," says Perrin, plainly.

"If it were moving, it wouldn't be posing, would it?"

"That must be Moogh logic," says Perrin. "Alright, I've had enough. Let's go." He turns and makes to walk away.

"What do you mean, go? You've only just arrived."

"I've seen what it does, which isn't that much, by the way."

"But you haven't sensed it," Maggie protests.

"What does that mean?"

"You haven't got present to the Moogh."

"What are you even talking about?"

"You haven't allowed it to touch you emotionally."

"Well, how long does it take?"

"You have to be open to it. Let it in."

"Yeah, whatever. I'll try again."

Perrin turns to face the Moogh and just stands there observing it patiently. He thinks that it looks like someone has dressed a statue in a bear suit. The funny, juvenile looking face is at both endearing and somewhat ridiculous. Perrin understands how the Moogh has a following mainly of women. It looks thoroughly peaceful, safe. And dead boring at the same time – like someone else's baby photos.

"Alright, I'm out of here," he says after another minute, standing in the wet grass watching the immobilized primate surrounded by birds and insects. Perrin starts to move, but Maggie reacts instinctively, putting out her hand.

"You haven't given it a chance," she says, anxiously, disbelieving that the Moogh's magic has no effect on Perrin.

"You know, Paprika, I'm a simple man. I like things I can touch. A brick that was thrown at someone's head, for instance. Or a wrecked car with blood on the seat. I'm not into this spiritual, voodoo bullshit. Sorry mate, I'm gone."

"But you wanted to come here," she protests.

"I didn't want to see the Moogh."

Maggie feels like she must have missed something important, along the way. "What?" she stammers.

"I didn't come here for the Moogh. I came here for the idiot people who get themselves hurt around the Moogh. I'm going to check in and get my gear ready for reporting real news."

Maggie glares at Perrin's back as he walks away across the grass towards the security gate. She is stunned, feeling the heat rising around her neck. She makes an exasperated gasp and tries to reconcile the nature of the emotion she is feeling. It's not hate, but a sense of intense frustration. She doesn't wish him ill. She just wants him to go away and never come back again.

Maggie looks back to the Moogh hoping for inspiration. However, for the first time in years, she doesn't get one. Turning again, she sees Perrin marching away followed by the two guards in their white catsuits with stun guns. "Do him now," she thinks. "Do him now."

UNMA Cocktail Party

After she leaves the Moogh, Maggie returns to the hotel by taxi. Inside the lobby, there is a large contingent of people checking in. It is an immaculately dressed international crowd, with the appearance of government officials, business people and some beautiful people thrown in for good measure. Maggie watches them for a while trying to make sense of who they are and what they are doing in the Moogh Zone.

Maxine is in the group, and when she spies Maggie, she comes over and directs her towards a lounge setting.

"How did you go with Perrin and the Moogh," Maxine asks. "Does he keep his security pass?"

"He's not that interested in the Moogh, actually," Maggie says, still stunned that such a thing could be possible.

"Well, there's a blessing."

"What's going on here?" Maggie nods towards the people at the reception.

"That is part of a group that UNMA has dumped on us at short notice." Maxine pronounces the acronym for United Nations Moogh Affairs with the word 'Unmarr'.

"Who are they?"

"Well, there's UN Bureaucrats, politicians, investment bankers, some movie stars and various beautiful people for shits and giggles."

Maggie starts to laugh aloud, and it feels good.

"Are you okay? You look like you're stressed," says Maxine.

"I'm okay. I just feel like I've been babysitting a gremlin."

"Just remember that the gremlin's Moogh privileges can be withdrawn with a single phone call."

"I understand."

"Something else," says Maxine. "Our UN friends are throwing a banquet upstairs tonight, and I didn't get to approve the invite list. So a lot of people who ought to be attending, yourself included, I'm sorry to say, didn't get an invite. I'm sorry about that."

"Well that's okay, I guess. You're going, though?" asks Maggie.

"Oh, you bet. If I weren't on the list, I would have bought the house down. Anyway, I'd better keep being the hostess with the mostest."

"I'll pass the message on to Perrin."

"Thanks." Maxine walks across the lobby, into the crowd of immaculate, and influential people. She seems to melt into the group like an ice cube in a glass of gin and tonic.

From the lobby, Maggie makes her way to Perrin's room. His door is open, and he is pottering around inside. She knocks, unsure of how she is going to react to him.

"Hey, Paprika, come in," he says, enthusiastically.

She steps inside and sees that Perrin's bed has been transformed into a staging post for a bunch of technologies such as cameras and recording devices.

"What's all this?" she asks.

"This," says Perrin, proudly swooping his hand across the bed. "This is the latest in journalistic technology."

"I just use a pencil and a notebook," Maggie smirking.

"Maybe that's why you don't trend."

"Really?"

"And maybe that's why I am here."

"Great."

"You can see how all these things fit together."

"Sure," she says, flatly.

"Can I show you around?"

"Why don't you show me around, Perrin. Then maybe I'll trend in the blogosphere. And then maybe you'll go away."

"You'd like that, wouldn't you?"

"I've enjoyed your company so far," she says, disingenuously. "But good things can't last forever."

"Well listen up then," Perrin lifts the first of the objects off the bed. "These are top of the range binoculars for seeing distant things as if they were close up. You could see a wart on the Moogh's ass from a mile away with these puppies."

"The Moogh doesn't have warts on its ass," says Maggie, conscious of the fact that she is smiling as she says this.

"Not yet it doesn't, Paprika, not yet," Perrin puts the binoculars down, and retrieves a pen.

"And what's this?" Maggie asks.

"This is a camera pen that writes upside down with an inbuilt LED light."

"That would be handy for taking pictures of something you have written on the ceiling at night," Maggie says dismissively.

Perrin looks at her oddly. "I guess so."

"Is that it?"

Perrin picks up a camera and says, "This is the *piece de resistance.* Wireless internet camera with built-in GPS and dictaphone. Press here, start talking. Press here, click, click, click, take photos. All automatically uploaded to the server with GPS coordinates. You remember those scenes in the movies where the bad guys destroy the reporter's film? It won't happen with this puppy. I'll give you a demo."

Perrin points the camera at Maggie and adjusts the focus. He says, "This is Paprika Tart, Senior Moogh Reporter, for the Fractious News Network. Live from the Moogh Zone." He fires off three frames and observes the back of the camera as a green light flashes and then goes blank. Then he retrieves his laptop computer and shows her where the photos and audio file have uploaded to the server.

"There they are, you see. And that little text file there has metadata including the GPS coordinates of where the photos were taken."

"You can delete those pictures now," Maggie says, flatly.

"You're right. You were scowling."

"I was not scowling."

"Whatever, you get the point." Perrin hits 'delete' and the files disappear off the server.

"You know that there's a UN contingent in town today?" Maggie tells him.

"That's all those fancy pants people downstairs?"

"Yes. Anyway, there is a function on tonight, and we're not invited."

"Really?"

"That's straight from Maxine."

"It must be true, then."

"So I'll leave you with your toys."

"And what are you doing tonight?" Perrin asks.

"I am going to finish writing up a story,"

"A story? Something happened today."

"I saw a little girl holding a Mooghie Toy and her father didn't know what the Moogh was looking at."

Perrin starts chuckling, "That's really going to trend in the blogosphere."

Professionally Embarrassed

In the morning, Maggie is seated in the hotel restaurant sipping coffee after her breakfast. The restaurant is abuzz. Besides the usual guests, the diners include the UN Moogh Affairs delegates. They all look decidedly seedy from the cocktail party last night.

Maggie retrieves her smartphone and opens the Moogh App to check on the photos that usually get posted on the social media section after such events. There are dozens of photos showing a swish affair with a big buffet, a chocolate fountain and copious amounts of alcohol.

There is a picture of Maxine schmoozing UN delegates, the way that she does. There is a picture of a crowd in rapt attention listening to an old military man. Next to him, tall and angular, is another man in military uniform. Maggie recognizes them as the Big General and his Chief of Staff. She flicks through a few more of the photos, familiarizing herself with the diverse array of people who make their living from the Moogh.

There are dozens of photos arranged chronologically, and it is possible to see the progression of the evening into the night. In the early pictures, everyone looks formal and stiff. However, as the evening progresses, there are noticeable changes: the neckties are looser, the grins are bigger, there is more haughty laughter, and people are increasingly placing their hands on other people's shoulders.

Maggie wonders whether this might be the basis for an interesting story, about how the Moogh and some booze brings people together. Then she halts these thoughts and her jaw drops open. "What the f…"

In the sequence of photos, sometime around 10 pm, Perrin shows up at the party. One minute there's no sign of him and the next, he's in nearly every frame. He's doing that thing he does where he talks rapidly with various people and they explode in laughter, patting him on the back as if he is suddenly part of the family or the inner circle.

"*Ughhh*!" Maggie drops the smartphone on the table like it was charged with electricity. She glares around the restaurant at all the UN Moogh Affair staffers who are suddenly hiding a dirty secret from her.

"How the f**k?" she mutters through gritted teeth. How come he was invited to the party? What's going on? Then a dark thought comes to her, and she retrieves the smartphone from the table and continues searching the images.

It is 11 pm, and Perrin has his arm around a senior UN bureaucrat who seems decidedly worse for wear. The official is looking at the floor while Perrin, sharp-eyed and clearly in charge of his destiny, whispers in his ear. Another photo shows Perrin standing next to the Big General. This time, Perrin's mouth is shut, and the Chief of Staff is eyeing him mistrustfully.

And then! The picture she had been dreading!

It is past midnight now, and the party has thinned out. In the foreground of the picture is one of the beautiful people in a cocktail dress drinking from a glass of champagne. Immediately behind her stand Perrin and Maxine. It is an odd scene, Maxine has her hand on his shoulder and is talking directly into his ear. Perrin is intently concentrating, almost with a grimace as if he is trying to understand something complicated and important. Another photo, later in the sequence, shows Perrin and Maxine eyeing each other. It is almost as if they share something that no one else knows, and they are confirming that each is committed to keeping the secret.

Maggie scrolls to the end of the photo shoot. There are more pictures of Perrin stirring the pot. The very final image is at three in the morning. The hall is empty apart from hotel staff cleaning up. Perrin is standing at the buffet, feeding. He is holding a glass of champagne and stuffing his face with tiger prawns dipped in what looks like thousand island dressing.

Maggie is stunned. She lays the phone on the table, feeling as if her brain were short circuiting. "How could… Why… Who…" she is so incensed that she can't even think a proper question.

Then, just to throw fuel on the sparking wires inside her brain, Perrin enters the restaurant and wanders over to her table. He has rings under his eyes, most certainly from alcohol abuse.

"Hi, Paprika," he says, lightly. "Hey, what's the matter."

For a moment, Maggie is stunned and incapable of speech. She doesn't know how to communicate or what to say first. Eventually, it blurts out, "Did you have a good night?"

"I had an awesome night. Can I sit down?" he indicates the seat opposite.

"Oh, why don't you sit here instead?" Maggie stands and points to her seat. "Given that you have taken over all my other roles."

"Don't be melodramatic." Perrin sits across the table eyeing her with a knowing look. "You need to check your messages, Paprika."

"You didn't think to invite me to the cocktail party last night?"

"I did think to ask you," says Perrin, eyeing her, wearily.

"Well, it would have been nice if I'd received the invitation."

"Paprika, listen to me, you have not been betrayed." Perrin lays his hands on the table, palms upwards as if to demonstrate she could trust what he was about to say.

"This is how it went. I harangued the dude from Moogh Affairs to put you and me – you and me – on the invite list. As soon as we were on the list, I went to get the okay from Maxine, doing my politics, you know. She said that she was delighted that you were on the list, so I called you, but it rang out. So I texted you with the details, and I even came down and banged on your door. I couldn't raise you, so I left you a note."

Perrin glances at Maggie's purse on the table with the corner of a white envelope sticking out the top. He shakes his head and then swivels around in his seat and raises his hand to get the attention of a waitress.

Maggie growls, feeling outsmarted. She remembers sitting up in bed, writing on her laptop when Perrin's call came. She tossed the phone to the foot of the bed. When the text message came through it was out of reach, so she ignored it. And when the knock came at the door, Maggie was half asleep, and lay there wishing that whoever it was would just go away. In the morning, she saw the note, picked it up and shoved it in her purse. The purse that is on the table next to her with the white envelope sticking out the top.

Perrin orders his coffee and then looks at the envelope and then to Maggie. She takes the envelope and opens it. The note inside is hand written in a style as exaggerated as Perrin's voice. "Paprika! We're in. Cocktail party. You're on the list."

Maggie folds the letter and slides it back into the envelope. She mutters something under her breath. "You shouldn't have gone without me."

"I shouldn't have gone without you?" says Perrin, exasperated. "I did think about that for a second. But the Moogh Desk is on the verge of getting closed down and one way to keep it going is to find some news that gets eyes on the page."

Maggie softens, feeling exhausted at being so pent up. "And how are you going with that?"

"Well, I learned that some poor mug who lives near here lost his house gambling Moogh Coin through the app. Says he wants to blow up moogh.org. I am going to interview him. What are you doing today?"

"I am going to follow up on the scientists who were looking for heavy metals in the Moogh fur," says Maggie, thinking that Perrin's story is infinitely more interesting than hers.

"Well, let's hope that they have made a groundbreaking discovery. We might be out of the woods then."

Plane Crash, Not

Later that evening, Maggie is in the bar in the Plush Hotel flipping through the wine list. She is still uneasy about having missed the cocktail party, but all the UN Moogh Affairs people have left town now, so the issue is no longer pressing. The interview with the scientists had gone well, although they had no new information to give her.

She runs her finger along the list not really concentrating on the words. She is distracted by an unsettling sound, something like an out of tune trombone. She looks around to see that Perrin is seated a little way from her. He has the camera lying on the table and he is concentrating on his smartphone. One of the bartenders places a fresh drink in front of him, and he looks up and starts talking. He's had a few and is probably still drunk from guzzling champagne until three am. He is over-talking, using too many words, too loudly. He picks up on a conversation with the bartender.

Perrin says, "I mean take a look at these mugs going all gooey-eyed over the Moogh. It's nothing but a sack of mangy fur and it doesn't even bloody move half the time. It's not like it can fricking dance! Talk about triple-threat. Not!" He laughs out loud and takes a sip of his fresh scotch. Then he starts again.

"I mean how do you know that it hasn't even died already. And all these so-called empaths. Empathy, my ass. If they had any empathy, they'd have it for the poor bastards who got sucked into paying two hundred and thirty dollars for a Moogh Coin and then losing it in an online casino. It's totally sewn up mate. They got everyone thinking that the Moogh is some planet saving superhero. When in reality, it's nothing but a monkey with a squashed head. Suckers."

"Squashed monkey," thinks Maggie, shaking her head

remorsefully. She remembers back to when he made a promise never to insult the Moogh in her company. Well, he is in her company now, whether he knows it or not, and he needs to pay the price. The business with the cocktail party, she is prepared to let that go. But the squashed monkey, oh no! Time for revenge.

One of the bartenders is standing patiently, waiting for her to place an order. He's a young guy, in his late twenties and looks like a bit of a Romeo. Maggie turns it on for him. It's a thing that she does with her eyes as she runs fingers through her hair. She makes a provocative look that never fails, for Maggie can go from beauteous to seductress within the wink of an eye.

"Hey," she says, with a mousy tone. She cocks a tiny grin with one side of her mouth.

"Well, Hi," says the bartender, the prospect of love setting him onto autopilot.

"Will you do something for me, that I can't do by myself?" Maggie asks.

"I'd sure like to try that," the bartender leans close, conspiring.

"You see that guy over there?"

"That muppet can't shut up. He's been here for hours."

"He's a journalist," Maggie confides, "You need to go over there and convince him that there's been a plane crash near the airport."

"Really? Has there?"

"No."

"Well that's even better, then, I guess. What for?"

"Would you pleasure me?"

"I'd love to pleasure you, honey." The bartender straightens up. The tip of his tongue seems to protrude from the side of his mouth. Then he moves towards Perrin.

Maggie retrieves her smartphone and quickly types a text message to Perrin. She looks up to see the bartender address Perrin. Maggie presses send and watches Perrin pick his phone off the wooden bar and read her message: 'moogh plane crash airport'.

The effect is electric. Perrin leaps from his seat. He skulls the scotch, grabs his bag and swoops towards the door. As he rushes past, Maggie calls his name. He halts and says impatiently, "What?"

"Hurry!"

Perrin grunts and flies out of the door. Maggie starts to laugh out loud. She taps the drink list on the table tip and watches the young bartender swagger over. He's coming to claim his prize, but he's not going to get anywhere near her.

"How did I do?" he asks, known full well that he duxed it.

"Oh, you did well," says Maggie. "I am going to have a glass of wine, and I am going to buy you one, too."

Moogh Ninjas

The trip the airport proves to be totally useless for Perrin, with no sign or word of any plane crash. He knows that he has been set up but rather than being angry towards Maggie, he feels a sense of having met his match.

He's concerned that he has been around the Moogh for nearly twenty-four hours and has nothing to report on, except for a broke and angry gambler. That bugs him not only because it will look bad on his part but because it also jeopardizes the very existence of the Moogh Desk. If he is going to keep Maggie in her job, he is going to need to perform some media magic tonight.

As he rides back towards the Moogh, he considers all that he has learned in the Moogh Zone. Specifically, he thinks through the pressure points that could form the basis of a major scoop. He knows that sometimes when the news isn't forthcoming, it is necessary to encourage its creation.

Back in the Moogh Zone, Perrin walks along the perimeter fence to see who's around. He comes across a young guy lounging around. He's a Flyby, one of the day trippers, and he looks bored.

Perrin watches as the guy drains the last drops from a beer can, crush it and toss it over the fence into the field. This behavior indicates a noticeable contempt for the Moogh - a man after his own heart! Perrin moves over and asks, "What do you reckon about the old Moogh then, mate?"

The guy sizes Perrin up and is quickly convinced that they are both of the same mind on the issue. "Its just some stupid fat guy dressed up in a monkey suit. They charged me three hundred and fifty bucks to get in here, hey."

"Did you get value for money," asks Perrin.

"It's a rip-off! I didn't even get a fricking show bag."

"Cracker!" says Perrin, laughing aloud. "What do you

reckon it's got in its hand?"

"No idea, mate. No one knows."

"If you had to guess, what would it be?"

"I don't know. Maybe a diamond or something."

"So how about we go get that diamond to put in your show bag."

"I didn't get a show bag. I just told you that."

"Dude," says Perrin, leaning close. "If you had that diamond, you wouldn't need a show bag. You'd just slip it in your pocket and bugger off."

"But what if it's not a diamond? It could just be a lump of monkey shit."

"Well, you'd have to find out, wouldn't you?"

"Find out?"

"You'd get your value for money if you were the guy who answered that question: is it a diamond or monkey shit, wouldn't you?" Perrin asks.

"I guess," says the guy, now interested. "That would make up for the three hundred and fifty bucks. What do I do?"

"The first thing is to let me take your photo. And second, we figure how to bust into through security."

"No way, dude, there's ninjas in there."

"Never heard of them, they don't exist," says Perrin dismissively. He raises his camera and snaps a photo of the guy. Then he retrieves his notebook and pen, "What's your name, mate?"

"Thomas."

"That's a good solid name. Thomas what?"

"How did you know."

"What?"

"That's right."

"Now I'm lost," says Perrin, lowering his pen.

"My surname is Watt."

Perrin laughs, "Funny. Alright, Watt, stand by to get famous. Come on."

They start walking towards the distant security gate, and Perrin outlines his plan. "I'll distract the guard, and then you run in as fast as you can and get the thing out of Mr. Moogh's hand. Then I'll photograph you."

"That's a stupid idea, they'll take your camera and film."

"Not a problem, dude. The camera automatically uploads pictures to the internet. You'll get the glory, I'll get the story."

"Plus there's Moogh Ninjas," Watt says, axiously.

"They don't exist," Perrin snaps, dismissively.

"You already said that, and you also said that you'd never heard of them."

"So what are they?"

"They're like these ninjas that protect the Moogh."

"*Pah*!" Perrin blows a raspberry. Mate, the worst thing that can happen is the stun gun. But then I'll get the pictures, and you'll get compo."

"Really?" asks Watt, suddenly interested, "How much?"

"Shitloads mate. Now, here we are." Perrin pulls up next to the security gate and peers inside.

"It's simple as a pimple," he says. "Wait until they are distracted then go for it."

Thomas Watt starts to snigger and holds his hands against his mouth. Perrin glares at him. "Dude, this is no time for monkeying around. This is serious Moogh shit. Are you good to go, or what?"

"Yeah, yeah, yeah. Whatever."

"Put your game face on, Watt," Perrin growls.

"Alright, I am good for it."

"Don't screw this up, mate. This is front page material. Zero to hero, you understand?"

"I am all good, mate. Good as gold."

"Okay, I'm going in." Perrin slaps Watt firmly on the shoulder and then wanders nonchalantly into the shelter that forms the security gate. He addresses the Level-4 security guard who approaches him.

"Hey babe," he says and raises his ID badge. "I tried to get an NOTAM through the Moogh App but couldn't figure out the code." He raises his smartphone for the security guard to see.

With the guard distracted, Thomas Watt dashes through the security gate and runs as fast as he can towards the Moogh. There is a flurry of commotion and shouting as the guards wise up to what is happening. Instantly, there are three Level-4 guards, foxy women in white catsuits crossed with red cork, giving chase.

"Hell, yes!" shouts Perrin and runs behind them, excitedly, clutching his camera.

Thomas Watt is a quick runner, and he heads directly towards the Moogh. It is fast and exciting, five people storming across the field, the Moogh several hundred meters away.

"Halt or be stunned!" shouts one of the Level-4.

Thomas Watt reaches the Moogh and dives for its right hand. He grips the Moogh's fingers and tries to pry them open. The Moogh barely seems to notice Watt's intrusion, just standing there calmly in its pose.

The guards are close behind, but they halt about ten meters from the Moogh. It is as if they have been instructed not to get too close.

Perrin halts a few paces behind the Level-4. He raises his camera and fires off a sequence of shots. He captures the scene of the drunk and brave Thomas Watt wrestling with the immobile Moogh. White camera flash illuminates the struggle.

Then suddenly there is another flash of light and a terrific noise close to Perrin's head. A gunshot rings out just a few feet away.

CRACK!!!

The percussion rattles Perrin's brains, and he falls to his knees holding his hands against the side of his head. His ear drum aches, and there are loud noises reverberating through his brain.

Disoriented, he turns to see a person standing beside him holding a rifle in the firing position. At first it is hard to identify whether it is a male or female because green fibrous camouflage covers their body. A security guard's torch beam flashes past and Perrin gets a brief glimpse to see that the person with the gun is a young woman. She has mud smeared on her face and intent eyes checking that her shot was true.

Perrin turns towards Thomas Watt. He's dead, slumped on the ground in front of the Moogh.

Perrin gasps, feeling his stomach rise in his throat. He looks back to the woman who fired the gun, but she's gone. In her place is a shallow depression in the ground covered with foliage, the foxhole in which she had served her shift watching over the Moogh, armed with a rifle.

Thomas Watt was right. There are Moogh Ninjas!

Perrin feels a wave of adrenaline wash through him, and he is alerted to the opportunity of the story. Human tragedy has unfolded, and he is on the scene.

He fires off a series of photos of the foxhole, illuminating the ground with white camera flash. Then he turns his camera towards the Moogh and marches forward, hitting the shutter repeatedly as he approaches. The Moogh and the dead Thomas Watt are illuminated every half second by the camera flash.

Suddenly, in front of Perrin rises another camouflaged figure, springing up out of nowhere. It's another Moogh Ninja – or maybe the same one. She pushes the muzzle of a pistol roughly against Perrin's forehead and he stops dead in his tracks. Camouflage hides the Moogh Ninja's face but Perrin can see that the hand is that of a young woman. He feels a sense of imminent death and does not resist when the Moogh Ninja pulls the camera from his hands.

Then the Moogh Ninja moves forward, pushing Perrin backward by the gun barrel. The pain in his forehead is intense and a wave of terror washes over him.

"Boo!" says the Moogh Ninja.

Perrin freaks-out. He turns and runs at full pace towards the gate. He stumbles and trips on the dark grass as he moves desperately out of range of the deadly, camouflaged women.

Through the gate, he runs down the road into the forest, and he keeps running until he is exhausted. He halts, bent over, holding his side, panting heavily. He looks around anxiously to see if he is being followed.

Then he moves into the trees and looks for somewhere to hold up while he catches his breath. He finds a fallen tree and lies down in the hollow underneath it. He inhales and exhales deeply, seeking to catch his breath without making too much noise. He peers out of his damp hiding place wondering how many Moogh Ninjas there are watching him.

He also wonders whether the internet uplink on his camera worked, and his pictures are safe in the cloud. Another Moogh death, this one caught on camera.

"Scoop," he mutters to himself under his breath, barely able to contain his excitement. "A real scoop."

Perrin Bound

The next morning, Maggie is padding around her room in her bathrobe when she hears a firm knock at her door. She pulls the door open as far the chain will allow and sees Maxine in the hallway, wearing a terse look. Maggie removes the chain.

"Hi, Maggie," Maxine's tone is flat.

"Is everything okay?" Maggie looks along the hallway, seeking some context to Maxine's unusual mood. At the end of the hallway, standing next to the elevators is a tough-looking woman in motorcycle leathers and a tattoo on her face.

"Unfortunately, everything is not okay," Maxine tells her. "Would you come with me?"

"*Ummm*, I'm not dressed for it right now."

"You are dressed just fine, would you please come?" Maxine takes a step aside and extends her hand towards the elevator.

Maggie looks along the hallway to try and learn more about what is bugging Maxine. It is unlike her to be anything other than bullish and jovial. But here she is, first thing in the morning, sour as lemon juice.

"I need to get my key." Maggie retrieves her key and slips her feet into hotel slippers. "Is everything okay?" she asks as she steps into the hallway and pulls the door closed behind her.

Maxine walks silently towards the elevator. The surly looking woman occupies almost the entirety of the small space. Inside, Maggie is affected by Maxine's demeanor, and she feels vulnerable wearing just her robe. She takes on a worried look.

"It's okay," Maxine says. "You are not in trouble."

"Perrin is, though?" asks Maggie.

Maxine confirms this by looking at the floor of the elevator and nodding somberly. The elevator door opens onto the basement garage. It is dim and gray and smells of cement and vehicle exhaust.

Maggie instantly feels self-conscious, thinking that she is going to get her hotel slippers dirty with oil. Nonetheless, she follows Maxine to the far end of the basement where there is a black electric Mercedes parked in a dark corner. Maggie is immediately put on edge when she observes three more of the burly looking women standing around the car, made from the same mold as the one in the elevator.

Maggie feels a chill run through her body, and she wonders what Maxine would have these women do that her Level-4 security can't or won't do?

Maxine retrieves her smartphone and activates the web browser to the Fractious News website. She passes the phone to Maggie, "This story went up a few hours ago."

Maggie takes the phone and reviews the story. It shows a photo of a Thomas Watt dead at the feet of the Moogh. Maggie gasps.

"Read the first two lines aloud," Maxine instructs.

Maggie complies, "A man was killed today in another violent incident in the Moogh Zone. Moogh correspondent, Perrin Speer, who witnessed the killing, said it was carried out by an all-female secret protective service called Moogh Ninjas." Maggie looks up at Maxine, nodding her head gravely.

"The Moogh Ninjas were a carefully guarded secret," Maxine says. "Now Perrin has told the world about them. This has compromised the Moogh's security."

"I understand." Maggie suddenly feels very cold and alone in the stark, dungeon-like basement. A chill runs through her as she wonders why she has been brought to this place.

"My committee wants Perrin disappeared."

"Disappeared?" asks Maggie.

One of the biker women lifts the boot of the electric

Mercedes. She raises a torch and shines it into the trunk. Maxine motions for Maggie to look inside.

Shiny black plastic lines the boot, securely taped into place. Perrin is inside, curled up in the fetal position. His face is bloody, and one of his eyes is swollen closed. His hands are bound tightly together with gray insulation tape, fists strapped together as though he were praying for his life. There is tape across his mouth and around his ankles. A white rope joins the wrist and ankle bindings together.

Perrin is completely immobilized, and the only part of him that moves is his uninjured eye. The eye swivels back and forth trying to see through the glare of the torch light and assess what awful thing will happen to him next.

Maggie gasps, feeling a complex mix of anxiety and empathy for Perrin's suffering. Most poignantly, she realizes that he wouldn't have busted into the Moogh Zone had she not taken him there in the first place. It comes back to her that he had shown no interest in the Moogh *per se*, and that he was interested only in the interactions of people around it. Further, she had been running interference on him doing his job and thus ramping up the pressure on him to perform. At the heart of it all is that she has been disbelieving that anyone could be unaffected by the Moogh. She realizes that her empathy is selective, only for people who share her views. She has mismanaged Perrin from the start and now a person is dead and Perrin maybe on his way to the same place.

She shifts her eyes from Perrin and sees that Maxine is speaking. Maggie asks, "What?"

"I said that normally, he would be shot dead and dissolved in bubonic acid so that all that remained was the bullet," Maxine says, plainly. "Do you have any other ideas?"

"Yes," gasps Maggie, tears welling. "Let him go."

"Let him go, just like that?"

"Yes," she says weakly.

"This calls into question the differences in the way that you and I view the world, Maggie Tarp," Maxine says, coldly.

Maggie feels her knees weaken as a little convulsion passes through her. It is as though something that she had thought of as safe has suddenly demonstrated a capacity for violence. Like a family dog baring its teeth. Maxine suddenly transcends from a tall and powerful protective force to a vast and terrifying threat. Maggie tries to speak, but words don't form.

"How do we ensure that Perrin won't cross our red lines?"

"I'll talk to him."

"You'll talk to him," Maxine chuckles. "Your words suddenly have influence, do they? There are no more chances, Maggie. One more slip up and he is gone."

"I understand."

Maxine thinks it through and then nods to her biker henchwoman. They lean into the boot, lift Perrin out and lay him on the concrete floor. One of the tough women retrieves a short curved knife. Perrin grimaces and shows a look of horror as the knife approaches, and then relief as the plastic is cut from his wrists. The woman takes hold of the tape on Perrin's mouth and yanks it hard. It makes a horrid fleshy noise as it comes free, and Perrin starts breathing heavily, he looks to Maggie seeming to plead for her help.

"I don't know anything about this," she says.

Perrin nods excessively, squirming on the floor.

"You can't mess around with these people, Perrin. I won't be able to help you next time."

Perrin keeps nodding, breathing heavily.

"My empathy for you is exhausted," she says.

"If anyone asks what happened to him," says Maxine, "he got mugged by the Moogh Underground."

She steps into the front seat of the electric Mercedes, and

her heavies follow her. The vehicle departs with a silent whine of the electric motor.

Standing, stunned, in the dark car park wearing a bathrobe and hotel slippers, Maggie watches the vehicle go. Then she turns her attention to Perrin, who is struggling to get to his feet. He grimaces, seeming to be wracked in pain all over.

Maggie helps Perrin stand. He is cramped and has cracked ribs, so he winces as he rises. She takes his hand and helps him to towards the elevator. He shuffles alongside gasping in pain. Inside the elevator, he props himself against the wall, exhausted.

Maggie presses the button and observes through the reflective surface that she has blood on her bathrobe.

"It's better to be cruel than be cruelled to," Perrin rasps. "Better yet, an end to the cruelty."

"What is that?"

"It's a poem that I made up while they were torturing me."

"I am going to get a doctor to see you," Maggie says.

"I don't need that."

"You are going to get it, anyway," says Maggie, firmly. "You need to stop making decisions for a little while, Perrin. Let me make them for you."

Perrin nods his head slowly and looks at the floor.

Moogh Zone Without Perrin

A week passes with Maggie in a strange headspace. Perrin is nowhere to be seen, holed up in his hotel room recovering from his wounds. It now seems strange that given all the effort she had put into getting him out of her life, she almost misses him. Plus, she is still ruminating about how her actions had contributed to him having got hurt. It is almost as though she had given him the rope which to hang himself.

If there is anything good to come out of this, it's that the Dim Director is effusive with praise for the Moogh Desk. The Moogh Ninja story has raised the network's ratings significantly. Maggie is receiving daily emails from the Dim Director singing her praises and big-noting himself on his 'augment-alation' plan.

In the afternoon, Maggie follows a group of scientists inside the fence and walks alongside the Moogh as it ambles along a forest road. Being present to the Moogh, she finds herself in a comfortable place and maps out a story in her mind about how the Adherents have adopted a form of yoga based around the Moogh poses.

That definitely won't trend in the blogosphere, she thinks. So she starts to consider how to make that story compelling and newsworthy. But she has no inner mongrel to draw upon, and the inner poodle just doesn't cut it, as the Dim Director had said.

She smiles as she thinks these things and realizes that her time as Senior Moogh Reporter with Fractious News might not last for long. Maybe she could get a job with moogh.org, she thinks. She considers it strange that Maxine has never mentioned this opportunity, given that she was very much aware of Maggie's situation.

Then she wonders whether she wants to work under Maxine, particularly after the nasty business of late. She thinks that moogh.org's behavior was unnecessarily violent.

Not only did they shoot dead Thomas Watt when they could have otherwise just arrested him, but they also tortured poor Perrin. The doctor's report showed that he had been very badly treated at the hands of the supposed pacifist organization. Cracked ribs, bruised kidneys, busted face. All evidence of a severe beating and 'stomping', whatever that was. And for what purpose?

Maggie thinks these things as she ambles next to the most peaceful animal on Earth, an animal that is either moving slowly or not at all. An animal that doesn't even eat. An animal that gets its energy from a mutually beneficial relationship with tiny plants. She watches the scientists going about their business, conducting their experiments.

For the first time in many years, the whole approach to the Moogh comes into question. Why can't you all just leave it alone, she wants to know. Why is that everything has to be manipulated and become a profit center for someone else.

Maggie's thoughts are disturbed as she hears a single word directed towards her, "Hey." Maggie sees Maxine walking alongside her, "How are you kiddo?"

"Kiddo?" says Maggie, as though she had never heard Maxine use the term before.

"How's our beautiful animal today?"

"He's ambling peacefully," she says glumly.

"I hope that this business with Perrin doesn't get you sent home," Maxine says cryptically.

Maggie looks towards her and sees the side of Maxine's face as she is looking towards the Moogh. A week before, Maggie would have read this as 'I hope that you don't get sent home by Fractious News'. But today Maggie senses another meaning, a veiled threat. She finds herself nodding gently, realizing that there are too many people with the power to send her home. Suddenly, she feels powerless and wonders how she has managed to survive this long.

"I could always get a job as an Adherent," she says, testing the water.

"Then you'd be working for me," says Maxine.

Maggie tests further, looking for something, not knowing what. "Although living without money would be a struggle." Maybe it is a coy way of teasing out whether Maxine would throw her a lifeline. A way of asking for a job without asking. Maybe it's the first time that she has set the tone and direction of the conversation with Maxine.

She looks at her feet pacing along the grass at the speed of the Moogh's amble, listening for a reply.

"Then we'd be poor together," Maxine says.

And that makes absolutely no sense to Maggie at all. Suddenly, she has a strange sensation. It is like a realization that she has had everything back-to-front or that she has only seen one side of everything. Her friend is an enemy. The truth is a lie. Violence accompanies peace. Her enemy is her friend… Maggie suddenly feels empty and cold. She desires to be in the presence of someone who cares about her wellbeing. Someone who speaks the truth.

"Come on," says Maxine, ushering her to walk faster. Normally, Maggie would comply without a second's thought. But today, her stubbornness rises to the fore. So is not going to come on. Quite the opposite, she turns and moves towards the security gate.

Professor Frederick Frotter

Back in the Plush Hotel, Maggie knocks on Perrin's door and waits patiently, anxious that he might have died or slipped into a coma. There's no answer so she pushes the door ajar. There is the sound of the TV playing, and Maggie cautiously enters the room calling out, "Hello!"

Perrin is sitting on the bed, with this laptop on his lap. A bad movie plays on the TV. He has a sunken, furtive look as he works intently on the computer. It seems as though he hasn't moved from that spot since the incident. His face lights up when he sees her, and he closes the lid of the computer.

Maggie sits on the end of the bed and Perrin turns the TV volume down to zero with the remote control. They look at each other impassively.

"How are you?" she asks, nervously.

"Better, now that you're here."

"Do you want to come back into the real world?"

"Oh, no thanks," Perrin says, shocked at the idea. "I'll stay in my fantasy land. It's much safer." He observes Maggie look down at her hands, pensively. "How is it out there?"

"It's not as simple as it used to be."

"Taking off the blinkers can hurt," Perrin says. He coughs and winces, grimacing as he clutches his chest.

Maggie shifts forward, anxious, wishing that there was something she could do. He has been tortured and abandoned, and she is somehow partly responsible. She feels compelled to speak, "Are you--"

He waves her away, gently. "It's okay. I got stomped by some gorillas. It'll take a while to heal."

Maggie looks at the floor, fearing that 'being stomped' has something to do with boots and the heavy women in biker leathers. She looks up, hopefully, "There's a presentation upstairs. It's Professor Frederick Frotter."

"I don't know what that means."

"It's fun. You should come."

"You mean I have to go into the real world?"

"Yes, you have to." Maggie stands and offers her hand.

Perrin thinks it through for a beat, realizing that his time in the health retreat is over. "Okay. I'll do it. Give me twenty minutes to shower and get dressed.

Maggie returns at the allotted time. Perrin is dressed and standing, looking just like the old Perrin, except that his movements are a bit slower and punctuated by the occasional gasp as an unhealed wound is tested.

They make their way to the auditorium and take their seats in the back row. Maggie scans the crowd to see who is there, seeing that it is largely the same people as the last time she was in the room.

Maxine enters and stands in the doorway observing the audience. Her gaze falls on Maggie and Perrin, and she moves over.

"Hi," she says lightly.

Maggie replies with just a smile, unable to speak.

"How's the Moogh Underground victim healing up?" Maxine asks coldly.

"I'm still alive. Thanks for that." Perrin says, cautiously.

"So, Perrin," Maxine says, tapping a fingernail on the table top in front of him. "Would you like to know where my name comes from?"

Perrin shakes his head hoping that she will either turn to ash or just go away. But that doesn't seem to be the right answer.

"No?" Maxine asks, ominously.

Perrin nods his head, and this elicits a more favorable response.

"Well, that's better." Maxine leans forward and says in a conspiratorial tone, "You see, my father was a Russian cosmonaut. He worked on the Proton rockets. He wanted me to be infallible, and he saw to it that I had all the proper education and training and a name that was fitting the task. My name, Maxine, is a portmanteau of 'maximum' and 'hydrazine'."

Perrin starts nodding again, anxiously, wishing it would end.

Maxine continues, "Hydrazine is a toxic and mutagenic rocket fuel. It's a hypergolic. I don't need a flame or catalyst, Perrin. I have an oxidizer inside."

Perrin's face goes white, and he stares at Maxine, nodding numbly. He gulps audibly and tries to say the word "okay" but he can only make a rasping noise.

"Anyway, enjoy the show. It's Professor Frederick Frotter." Maxine beams a big smile at Maggie and then departs towards the front of the auditorium.

"Are you okay?" Maggie asks.

"She is really scary, hey?"

At that moment, a strange looking man enters the room. He is old and scruffy with a long beard and unkempt yellow and white hair. He approaches the podium where one of the moogh.org staff is manning a laptop. The computer is connected to the projector that shines onto the white screen an image of the folders and files on the desktop of the computer.

As soon as Frotter lays his hands on the computer mouse pad, things start to fall apart quickly. The projector shows the cursor zoom this way and that as he tries to locate his power-point presentation. He avidly clicks every icon and button. Folders pop open and are then just as quickly abandoned or closed down. The cursor zooms this way and that. All the while, Frotter is grumbling, using the words "damned computers," and "where the hell is it?"

"Jeez, I'm getting dizzy," says Perrin, trying to keep up with the exaggerated movements on the screen.

Then Frotter presses the wrong button, and the computer shuts down. There is an audible moan from the people in the crowd. Then the noise of chatter rises.

"Oh, here it is," says the Professor, retrieving a memory stick from his pocket. He inserts the stick into the computer and presses keys on the keyboard, willing it to boot up faster. "Come on, come on," he exhorts. And then he says, "damned computer," again. He is shaking his head, and his frizzy yellow and white hair vibrates around him, seeming to endanger a bolt of static electricity.

Perrin starts laughing and wincing, holding his ribs. "This is really funny."

"Its always like this," says Maggie, smiling, "a blend of brilliance and ineptitude."

Eventually, the computer is up and the first slide of the presentation displays on the screen. Maxine steps forward to introduce the Professor, but he intervenes.

He waves her aside. "I'm Frederick Frotter from the Institute of Planetary Phenomena."

Frotter looks across to crowd to ensure that the people are paying attention. "Today, I am announcing a breakthrough in the scientific understanding of Moogh Tears."

"Wow." Maggie retrieves her hemp and organic cotton, Moogh embossed notebook and scribbles a spiral in a circle on the top of the page to gauge the sharpness of her apple wood pencil.

Frotter raises a document and holds it in the air. "This is a peer reviewed paper co-authored by our Institute and the Stockholm Resilience Centre. It describes our theory of the meaning of the Moogh Tears."

Frotter presses a key on the computer keyboard, and the next slide appears. It shows a diagram, a circle dissected by lines like a pizza with nine slices. Some of the slices are overlaid by colors, green, white and red. Around the outside of the diagram are words that name each of the individual slices.

"This is a diagram representing planetary boundaries. There are nine biophysical boundaries that cannot be exceeded if we humans are to stay within the safe operating space for humanity. We believe that this framework – Planetary Boundaries – explains the Moogh Tears. The Moogh cries to demonstrate that we have exceeded the boundaries."

"Wow." Maggie lays down fleeting, wispy strokes of the pencil as she creates her exquisite shorthand.

Frotter moves to the next slide. It shows two images of the Moogh standing in the 'my left shoelace is undone' pose. A dead bird is located a few inches from its toe. The pictures represent two separate instances and two different types of bird. In one image, the Moogh is crying and in the other it is not.

"Why would the Moogh cry over one bird and not the other?" the Professor asks, "The ornithologists tell us that this bird is a non-native to the region, an invasive species. And this bird, over which the Moogh is having a big cry, is a native and is an endangered species."

Frotter goes back to the slide with the diagram and uses a laser pointer to highlight the words 'biodiversity' of which there are two classifications 'functional' and 'genetic'. "The Moogh can identify species and their conservation status."

There is the sound of chatter from the audience. The chatter dies down when Frotter moves to another slide showing the Moogh in another pose, crying. The Moogh is looking up in the sky, tears pouring down its cheeks, its right hand catching and absorbing some of the fluid.

"Now this day we are scratching our beards a very long time," Frotter says, chuckling. "What was Mr. Mooghy crying at? Maybe nothing at alls. Maybe that naughty Moogh was making us do a wild swan chase. Then our friends at the European Space Agency have a solution. We cannot see because it is daylight, a Cygnus shuttle burning up in the mesosphere right where the Moogh is looking. It was a swan chase after all." Frotter starts chortling the way he does when he has made a joke. The people in the audience look at each other, confused, hoping that someone has figured it out.

A hand raises and Frotter points the laser pointer at them. "Oh, sorry," he says. "Your question?"

"What's a Cygnus shuttle?"

"A space vehicle used to take things to the International Space Station. Also used as a bin for astronaut rubbish."

"And why is it burning up?"

"Astronauts use the atmosphere as a waste incinerator."

"Why would they do that?"

"It is convenient for them." Frotter goes back to the original slide. He draws a green laser circle around one of the sections: Novel Entities. "Novel Entities are heavy metals, plastics floating in the sea and other man-made pollutants such as the smoke particles left behind when space debris burn-up. We believe that this is what the Moogh tears mean on this occasion."

Frotter advances to the next slide. It shows nine images of the Moogh in various poses, crying. Frotter describes each of the planetary boundaries that the Moogh is crying over. "The statistical analysis shows the Moogh tears correlate with the planetary boundaries at the 95 percent confidence interval. A question over here," says Frotter.

"Are you suggesting that the Moogh knows the work of the Stockholm Resilience Centre?"

Frotter starts laughing. "No, no, no. Mr. Mooghy has never ambled to Sweden. But the Stockholm Institute and Mr. Moogh have the same ideas about the tipping points of the Earth system."

"And where would the Moogh get this wisdom?"

"We have no answer for that," says the Professor. "Maybe Mr. Moogh is a Gestalt of Gaia, I heard that said once."

"Hey. That was me," says Maggie, excitedly. "Frotter quotes Tarp. Did you get all that?" she asks Perrin.

Perrin is intently checking his smartphone. "Its all a bit techno-snooze for me," he says. "But you know what he's talking about?"

"Of course. He's confirming what everyone knew already. The Moogh is a messenger of sustainability." She looks towards Perrin when she hears him make a whistle noise and shift forward, peering into the smartphone screen.

"You should check this out." He slides his smartphone over to Maggie. The Mooglemap is open, and the data shows that the Moogh has been ambling parallel to the coast but suddenly turned ninety degrees and is now ambling through the dunes, directly towards the beach.

Perrin runs his finger across the screen and activates the Moogh Messenger that shows little yellow triangles with exclamation marks inside. The text says that Moogh Security is panicking because they have been unable to erect fencing in the dunes, and there are hundreds of Adherents and FlyBys swarming the area. A red triangle appears with an exclamation mark inside. The text reads that Moogh Underground has arrived.

From the front of the room, there is a beeping noise. Maxine stands and walks rapidly up the aisle, her phone gripped in her hand and a terse look on her face. Perrin swivels the camera towards her and fires off a shot. "I smell conflict and drama." He shines a big grin at Maggie. "Coming?"

On a Beach

In the time it takes for Maggie and Perrin to reach the dunes, the drama has settled in. They stand at the top of the sandy mountains that offer a view over the long stretch of beach. The sun is high, the ocean cobalt blue and flat, just small waves rolling quietly on the shorelines.

The beach is swarming with people. There are hundreds, maybe thousands, on the beach, and it is completely out of control. Visible from the top of the dunes are the normal mix from the Moogh Zone. There are Adherents, Flybys and Nonsticks swarming the area. There are also mobs of Moogh Security trying – and failing – to keep order. And right in the middle of it all, the Moogh is ambling, nonchalant and with not a care in the world.

Perrin makes a particularly grim chortling noise and holds his ribs. "This is so cool. Maxine will be shitting herself," he says.

"Look," says Maggie, anxiously, pointing into the distance to where a phalanx of Moogh Underground, two hundred angry, drug-addled young men, are running along the beach. At the head of the group, one carries a Moogh Flag. It looks immaculate as it flutters crisply, the orange Moogh Orange material glowing brightly against the blue sky.

Perrin raises his camera, zooms in and fires off some shots. "Awesome," he mutters as he watches the Moogh Underground butt up against a squad of Level-4 Moogh Security. The stunning women in red-strapped, white cat-suits, strut forwards in unison, zapping the Moogh Underground with their Rhinox-25 stun guns. Their handiwork is plainly visible as dozens of young men are tossed unconscious onto the sand. Stunned and paralyzed, they writhe or lie completely still from the electric shock.

Perrin hits the button on the camera and records some headline words, "Stunning women, stunning men. Pop go the corkers."

On another battlefront, visible from the top of the dune, UNMA forces wade into the melee with rubber truncheons swinging. This multinational group is recognizable by the many colors of their skin and the orange and pale blue of their helmets.

Also on the beach, trying and failing to keep order, is a contingent of self-styled Adherent vigilantes. Normally, the Adherents are pacifists, but with the Moogh's security in such parlous shape, the white gloves are off.

There are shrieks of pain mingled with whoops of joy as the normally tight-knit order around the Moogh plays out as an uncontrolled riot. In amongst it all, the Moogh ambles along the beach, oblivious to the human drama unfolding all around.

A group of Adherents struggles to form a protective ring around the Moogh. They do this by locking their arms together, with their backs to the Moogh. Inside the crude human barrier, heavy set Adherents deal roughly with the Flybys who are getting too close to the Moogh.

While the Adherents are typically empaths, there is something distinctly un-empathetic about the way that they deal with the inebriated youth; it is elbows and fists at close quarters for these wayward boys.

One lad staggers around clutching his bloody face only to be trampled on by the ring of Adherents, keeping pace with the Moogh as it ambles, nonchalantly along the beach.

Perrin scans the beach, excitedly, like a hungry man assessing a buffet. He focuses on where the drama is most intense, equivalent to where the champagne, tiger prawns, and thousand island dressing would be.

Maggie, standing beside Perrin, is in quite a different emotional state. She's not excited. She's anxious, and her only interest is the well-being of the Moogh. She studies the Adherents in their circle; arms interlocked, a human shield around the Moogh. And just to show its complete disdain for the affairs of man, the Moogh has adopted Pose #8 Swan Dive.

The Moogh is facing the ocean; gazing at the little waves lapping the sandy beach.

"Alright!" says Perrin, slapping his hands together, excitedly. "I am going to cover the stunners." He glances around at Maggie. "You alright?"

"Just go," she says and watches as he moves ahead quickly, descending the sandy pathway between the dunes, careful not to aggravate his wounds.

She follows him and as soon as they reach the beach, they go their separate way in the crowd. Perrin turns and calls back to her, "Paprika. Be careful!"

All on her own again, Maggie finds a path through the madness to where the Adherents form their protective ring around the Moogh. She stands there, observing the Moogh, trying to intuit how it feels about all this chaos and what might happen next.

Another phalanx of Adherents arrives and forms a second ring around the Moogh, this one encompassing the first. Maggie finds herself trapped inside. She recognizes some of the Adherents and takes the opportunity to speak with those of them that she likes. She asks them how they feel. They tell her that they are terrified, afraid for the Moogh and yet unconcerned for their personal safety.

There is the sound of a whistle blowing, and Maggie turns to see the riot police arrive, following the protocols laid out in the Declaration of the Adherents. Tear gas canisters fly through the air, and there is the sound of automatic gunfire as the riot police fire over the heads of the rioters. Panic ensues, and there is a mad stampede to get off the beach. Hundreds of frightened people scramble up the sandy tracks between the dunes. Some fall and are trampled by those behind.

Then the Moogh breaks its pose and begins to amble again. Hundreds of Adherents forming the protective circles shuffle along the beach. The Moogh stops, pivots and then ambles directly towards the water's edge. The circle starts to come apart as the Adherents find themselves knee-deep in the sea with waves rolling against their thighs.

Maggie watches in fascination as the Moogh ambles onto the wet sand and then into the sea. It halts in water knee deep and then throws the Pose #2R My Shoe Lace is Undone, right foot.

The Adherents resume their protective barrier, now forming a semicircle around the Moogh with its seaward side unprotected. The Moogh holds this pose for only a little while and then starts to move again, seaward.

The Moogh takes a step forward, looking down at its feet. It seems to be fascinated by the seawater around its feet. It takes another step and another and slowly walks into the sea.

As the water rises to its belly, the Moogh leans forward and then it allows the water to hold it afloat. Then something more. The Moogh pushes forward its left arm and claws the water with its free hand, holding the clenched right hand clear of the water.

Walking a few paces towards the water, Maggie is stunned at what she sees. Never in years of following the Moogh had she even suspected that it might do anything other than amble and pose. And yet, here it is; the Moogh can swim. That's news!

Maggie suddenly feels journalistically inadequate. She has only a hemp and organic cotton notebook and apple-wood pencil at her disposal. She needs to get her hands on some serious journo-tech. She spins around, hoping to see Perrin snapping away with the wireless camera. But Perrin is not interested in the Moogh. He only cares about the human conflict that the Moogh engenders. Maggie feels lost, inadequate, confused.

Then the reality of the day presses back upon her. Amongst the Adherents, there are screams and howls and a fight breaks out. Someone shouts five words and these are said over and over again, gaining in volume and anxiety every time they are repeated.

"The Moogh is leaving us! The Moogh is leaving us!"

The message quickly spreads across the beach and up the dunes. Soon all the people who had fled the riot police return and the beach is again flooded with thousand of people overrunning the security forces. Some people reach the water's edge and collapse, tears pouring from their eyes. Others plunge into the water and swim out towards the Moogh. One person, unable to swim, starts to drown, and thrashes wildly, calling out.

A security helicopter swoops over the beach with the loudspeaker and the pilot bellowing, "Get out of the water! Get out of the water!"

Around the Moogh, there are suddenly shadows and fins slicing through the water. Someone else shouts, "Shark!" and the swimmers turn and race back to the beach in panic. It is not a shark, but a pod of Tangalooma dolphins accompanying the Moogh on its swim.

Maggie stumbles into the sea, stunned as she watches the Moogh move slowly away from the shore. She is aghast and barely aware of the salty water lapping against her legs. Around her, thousands of Adherents and Flybys flock to the water's edge, wailing and thrashing their arms.

In the distance, Maggie sees the sleek gray form of a warship cutting a hole in the horizon. She stares at this ominous structure, her mind in a frenzy trying to determine the meaning. Is it a threat to the Moogh or a protective force? She thinks back to the discussion with the Dim Director and the trending story about the Big General's plan to nuke the Moogh if ever it were to move outside of a national territory. She looks at the Moogh, slowly swimming with one hand out of the water. It is heading to a place beyond Spanish national waters and towards the high seas where the rules of the Peace Park do not apply.

Out there, there is no United Nations mandate to protect the Moogh.

The Volcano Erupts

A wave of anxiety washes over Maggie. On her forearms, her skin has turned to gooseflesh and the hairs are erect, as though they could protect her against what is to come next. Instantly, she is in action fighting her way through the crowd of wailing Adherents, struggling to get traction in the sand.

Then suddenly – as if from nowhere – a tremendous noise pounds across the beach blasting the air from her lungs.

BOOM!!!

The sound pulses through her stomach and Maggie hits the sand, terrified. Breathing heavily, she raises her head to look around and sees that almost everyone on the beach is similarly face down in the sand. The Flybys, Adherents, Non-sticks, Moogh Underground and even most of the security have all been floored by the blast. Noticeably, Perrin is still on his feet, moving around quickly, snapping pictures.

With the explosion is still ringing in her ears, Maggie looks around to see where the noise originated.

"Over there," someone shouts, pointing out to sea.

In the distance, a massive plume of gray smoke rises from a triangle shaped island. At the top of the triangle is a sputtering of bright orange blobs and flame. A volcano has blown the top off its caldera, and the shockwave has spread from the source of the explosion.

There is a trembling underfoot as seismic waves spread out from the epicenter of the eruption. Thousands of people are back on their feet now, staring in awe at this magnificent and terrifying sight.

There is a plume of hot, bright orange ash rising from the center of the volcano. It rises rapidly, propelled upwards by its intense heat. As it interacts with the air at altitude, the other edge cools and begins to fall, forming the distinctive mushroom cloud.

"It's a nuke!" shouts someone.

Maggie comes to her feet, stunned. "No," she says aloud. She looks out to sea and observes the Moogh continue its one armed swimming, as if nothing odd was going on. It wasn't a nuke. She needs to stop listening to the voices from the crowd.

There is a collective sound of awe as another vast belch of gas and ash shoots up through the center of the original plume. Maggie looks towards the volcano and sees something strange.

Between the volcano and the beach something is shimmering on the sea surface, a sort of white cloud that grows higher and whiter as the seconds pass. Is it getting bigger or closer? Closer! Another shockwave smashes over the beach with a terrifying noise throwing thousands of people flat on their faces again.

CRACK!

There is a mad scramble as a pack mentality kicks in and a stamped begins. Sand is kicked up in the air, people trip over and are trampled by others. Someone shouts "Tsunami!" and this only serves to drive the panic harder.

Maggie finds herself immobilized as the beach erupts with panicked people all around her. She glances around to check on the Moogh. It seems unaffected by these noises, slowly continuing its one-handed, overarm swimming motion in the direction of the distant warship.

Is the Moogh in danger as it swims towards the high seas where the Peace Park does not apply? Is the warship a friend or foe? What is she to do? She thinks fast. She needs to get to the Big General.

The General's Plan

Getting an audience with the Big General is not going to be easy, Maggie thinks. She starts making calls on her mobile as she trudges up the dunes. However, with moogh.org and Moogh HQ in a panic, no one is answering phones.

She recalls seeing the Chief of Staff at one of the recent presentations and thinks that she can probably get his number from the roster list from that event. She returns to her hotel and gets online with her laptop. She finds the mobile number and is relieved to get through straight away.

The Chief of Staff is not only polite to her; he seems to go out of his way to help her. Of course he can get her an audience with the Big General.

It is evening by the time Maggie arrives at the hotel where the Big General is staying, and the concierge has one of his staff escort her to an upper floor. Here is a function room in which the central feature is a buffet large enough to feed an army. The army, in this case, wears black ties and cocktail dresses. With her designer clothes, Maggie fits right in. The concierge directs her to the Chief of Staff. He is tall and handsome, dressed in a crisp uniform with colorful medals. He takes her hand, and she finds his grip to be surprisingly warm.

"I am Maggie Tarp, from Fractious News."

"Of course you are. The Bug Writer. I love your columns. Very empathic," says the Chief of Staff, grinning. "Come on." He leads the way and Maggie has to trot to keep up with his pace.

"Really?" says Maggie, thrown off balance by his comments. "What did you say?" she asks, "Bug Writer?"

"That's military parlance for the Moogh."

They arrive at a room outside of which are posted two very serious looking uniformed men with rifles. Maggie is suddenly flushed with nerves as the Chief of Staff ushers her into the room. It is a plush hotel suite, and there are a half dozen people interacting with the General.

The General is about sixty, heavy set and intense. While the other people in the room seem to have trouble getting and keeping the General's attention, the Chief of Staff gets through straight away.

"The Bug writer's here, Sir," he announces.

The General seems happy for the distraction, and he dismisses his entourage and ushers Maggie to a lounge setting.

"I'm Maggie Tarp," she says offering her hand. The General's grip feels like her hand is trapped in tank tracks, and Maggie wonders if she'll ever get her fingers back.

"You have a few minutes of my time, young lady," the General tells her. He rests back in the settee, seeming to become instantly relaxed.

"General, I understand that it is your intention to kill the Moogh when it reaches the high seas."

"You ought to check the field manual of military euphemisms," the General says, cryptically, leaving Maggie perplexed. She goes to ask another question but is stumped for what to say.

"Killing the Moogh is not our preferred terminology," says the Chief of Staff. "We do, however, have plans to neutralize the Moogh threat."

Maggie writes the words 'neutralize the Moogh threat' in her hemp and organic cotton seed notepad. "And how would you do that?"

"Well, we'd kill it, of course," says the General, laughing, "we just wouldn't use that word."

"I don't understand how you could view the Moogh as a threat?"

The General gets serious, "The Moogh is a trans-boundary disturbance to international order. It is a potential biohazard and possibly the source of a terrifying new global pandemic."

"The Moogh Flu," nods the Chief of Staff, gravely.

"And if the Moogh falls into enemy hands, it could become weaponized," the General continues. "And I will not let my boys face an enemy armed with Moogh technology."

Maggie feels overwhelmed by this new information. On one hand, she is angry that these powerful men have taken a threatening posture against the pacifist Moogh. On the other hand, she is getting a real story, and that can only work in her favor. The other powerful man in her life, the Dim Director, will be pleased with her, she thinks. And so will Perrin.

Maggie asks, "Is it true that the military plans to deploy a nuclear weapon against the Moogh?"

The Big General leans forward, excited to share his news. "We have at our disposal, a vast and diverse array of modern weapon systems. A 300 kiloton W61 nuke deployed from a Tornado is just one option to neutralize the Moogh threat. However..." The Big General raises his finger for emphasis. "Given the prohibition on the use of tactical nuclear weapons, the nuke is not our preferred weapon."

"Are you concerned about radioactive fallout?" Maggie asks.

The General looks to the Chief of Staff and laughs, "Nah. The fish love it."

"And when would you look to undertake this mission?"

"As soon as the Bug is on the high seas."

Maggie's heart rate quickens. "When do you anticipate that to be?"

"That's about two days from now at the rate it is swimming," says the Big General. "Now if you don't mind lady, I've got a military to run. There is a lot of bad guys out there who need neutralizing."

The General stands and extends his hand. Maggie looks at it mistrustfully, hoping that he won't hurt her. She shakes and he takes a step away from her and is immediately surrounded by his entourage.

Maggie feels deflated. Everything that she feared has been reaffirmed. She has confirmed that the General plans to kill the Moogh, but she doesn't have any new information except that he no longer intends to do it with a nuclear bomb. That story is hardly going to trend in the blogosphere. She feels as though she has wasted the interview. What would Perrin have done in this instance?

The Chief of Staff beckons her, and she walks with him out of the hotel room. In the hallway, she addresses the Chief of Staff.

"If the Moogh as has potential bioweapon capability, it seems odd that you would want to kill it."

The Chief of Staff glances at her as he walks on silently. Maggie thinks that she is onto something and tries to think like Perrin for a while. What would he do? He'd be annoying and keep pressing the point until he got a reaction, and then he'd lever off that for more. Maggie tries to emulate Perrin.

"Why would the military want to destroy the Moogh if it believes it could foster a major new weapon system?" Maggie asks, "I mean, if I were running the military, I'd want that thing safe and in our hands so that we could study it."

Unfortunately, she doesn't have Perrin's annoying persistence. She is simply too pleasant, even when she is trying to be annoying. They enter the convention room and pass through the schmoozing military folk to the other door with the Chief of Staff remaining silent.

In the doorway, he turns to her with a warm smile. He shakes her hand and agrees, "That does seem contradictory, doesn't it? Goodbye ma'am."

Self Immolation

After the interview with the Big General, Maggie returns to Moogh HQ to catch up with Perrin. On the way, she acknowledges her new found respect for his journalistic skills. She even thinks that she was remiss for not taking him with her to meet the military folk. The is no way that Perrin would have come away without a scoop.

As she enters Moogh HQ, she halts in front of a big TV screen and watches a news bulletin covering the volcanic eruption. The screen shows a most extraordinary image of the plume, amplified in awesomeness by the color of the sunset.

The image shows a wide landscape in which the central feature is the triangle shaped mountain where the volcano's caldera lies. From the center of the volcano is a column of ash that raises and then spreads in an archetypal mushroom cloud shape. The cloud is massive and bright orange. And its surface seems to be animated, rippling and bulging as though it were part of a living organ, a giant tumor in the sky.

In the background, the evening sky is cobalt blue and the bright orange of the ash cloud clashes terribly, the garish colors of nature. The upper part of the cloud butts against a high-level wind, and this causes the cloud to stop rising and instead to spread horizontally with its lower surface a perfectly flat plane. The flat underbelly of the cloud is patchy in color. Some parts are bright orange and in other places, the blue sky beyond can be seen. This extraordinary sight seems to cover hundreds of square miles, dominating the skyscape.

Maggie sees one of the scientists moving past her, and she interrupts him and asks, "Is this anything to be worried about?"

"That depends."

"Depends on what?"

"Well, the volcano is a natural event and nature has coping mechanisms, so the planet is fine. The other question is whether the humans will be harmed. Probably the biggest concern is the ash getting into the stratosphere. If that happens, it could contribute to global dimming and global warming."

"Has it?" asked Maggie.

"We don't know yet. You see this high-level wind interfering with the cloud, flattening it out?" The scientist points to the flat surface at the top of the cloud on the TV screen. He says, "That suggests a stratospheric jet stream wind. So if I were a betting man, I'd say that the volcano is pumping ash into the stratosphere."

"Hmm," says Maggie, "Thanks for that."

"Say," did you hear that Kathmandu got flattened by an earthquake," the scientist tells her.

"I didn't know that."

"Exciting times, huh?"

Maggie wanders further into Moogh HQ and finds Perrin at a bench working on his laptop. "Hey, Paprika," he says, glancing up.

"What a day!" she slumps into a chair, suddenly exhausted. "Did you cover the stampede on the beach?"

"I got that. That's in. I got some awesome shots of this guy getting his head squashed in the sand. You want to see?" Perrin says excitedly.

Maggie waves her hand, "I'll trust it is up to your normal standard."

"And I love this volcano. People are freaking out all over the place. Have you seen the imagery?"

"I have."

"How did you go with the General?"

"He confirmed the story about killing the Moogh," Maggie says flatly, "but not with a nuke."

"Did they say why?"

"Public opinion."

"Pah! That's no story," says Perrin dismissively. "You can cover that one."

"I think that there is something else going on," says Maggie, thoughtfully. "The General says that he is concerned that the Moogh might be captured by an enemy of the US and could be weaponized."

Perrin starts chortling aloud, "That's funny."

"But if it has weapons value," Maggie continues, "wouldn't they rather capture it and develop it themselves than kill it?"

"The Moogh as a weapon, I don't think so," says Perrin. "Maybe if you dropped it on someone's head. Or used it to bore the enemy to death. That might work."

"Yeah," agrees Maggie, "No much of a story there. How many words did you submit on the stampede?"

Maggie waits for a reply, but none comes even though Perrin is just a few feet away. She looks towards him and observes that he has suddenly gone very still, as though he were picking up faint signals and didn't want a movement to interrupt them. Maggie is reminded of a hunting dog, motionless, listening for the sound of its prey. She shivers, feeling hairs rising on her arms for the second time in a day.

"Something's happening," says Perrin, quietly. He slowly turns his head towards the opening of the tent. In that instant, the door bursts open and a young guy rushes in, shouting, "There's been an accident!" He steps further into the tent, shouting desperately, "Help! Help!"

The young guy has fear in his eyes and Maggie instantly feels a pang of anxiety at the dark thought of someone suffering. She stands, but an inner force prevents her from doing anything else. She remains motionless.

Perrin's reaction is exactly the opposite: a spike of adrenalin and a quickening of pace. There's a story that is bleeding and pungent with the aroma of distress. He swoops the camera from the desk, moves to the door of the tent and scans the surroundings outside. There is smoke rising in the distance and the distinctive noise of people panicking.

"Paprika!" he yells, then swoops out the door.

Maggie responds immediately, grabbing her notebook and following Perrin outside. He is moving quickly towards tent city. Maggie has to speed up because Perrin moves fast. He darts left and right asking quick questions of people as he swoops past.

Perrin disappears from view into the crowd and Maggie halts, feeling as though she'd been deserted, and not confident to approach the site of the accident on her own. She stands, scanning the crowd, hoping to see something to relieve her discomfort.

There are disturbing noises coming from within the large crowd of Adherents. There are shouts, screams and other noises that are disturbingly out of place. Maggie's stomach clenches as she observes a pattern forming in the group up ahead.

Then suddenly, the crowd opens up, and a man staggers forward, burning. An intense orange flame billows from his body curling up around his throat and covering his face. He walks forward, pitifully shouting, "Ow! Ow! Ow!" and holding his arms away from his body as if the pain of being burnt alive was felt most intensely under the armpits.

"He's on fire!" shouts someone close to Maggie. This sound paralyzes her, the words reinforcing the horror of observing a man being burnt alive.

The burning man slows and comes to within ten meters of her. Then he drops to his knees with his arms still held out in front of him. He falls flat on his face on the grass, the flames still rising from his body. Black smoke curls into the air carrying the pungent odor of burnt flesh. His back shows the ugly pattern of skin that has blackened and curled.

A wave of terror washes over Maggie and a great

commotion rises from the crowd. Another voice rings in Maggie's ear as someone yells, "There's more!"

Suddenly, a cacophony of shrieking begins. Dozens of people start to move in an undisciplined fashion; rushing, bumping into each other. Panic sets in and Maggie sees the crowd open up, exposing a clearing in which there are four people raising buckets over their heads and drenching themselves with a translucent pink fluid. One of them strikes a cigarette lighter trying to excite a spark, and then he succeeds.

WHAFF!

There is the sound of the fuel igniting and orange flame bursting out from the self-immolators. Simultaneously, they begin their flaming zombie walk. People run around screaming, panicking, freaking out, and Maggie finds herself incapable of movement even though one of the burners is staggering right towards her.

Through a veil of flame, Maggie sees a the face of a young woman – a woman of her age. Her face is contorted with a look of agony and confusion, seeming to ask, "This hurts so much. Why is this happening to me?"

The young woman seems to take solace from Maggie, and she changes the direction of her flaming zombie stagger. With her arms, held up away from her body, it creates the impression that she wants to embrace Maggie. For a terrifying moment, Maggie realizes that this is to be her fate, but even the horror of being burnt alive in the arms of a stranger is unable to make her move. Just a few meters separate Maggie from this horrible furnace when the woman stops in her tracks. The burning woman starts to cry. She expels four or five bursts of anguish then falls flat on her face on the ground.

Maggie feels the heat of the flames on her bare legs, and she comes to her senses. Taking three steps backward, she stares, aghast, at the young woman dying in front of her.

The last thing that Maggie remembers of the self-immolations is the sensation of falling. She blacks out and collapses onto the grass.

Hospital Visitors

Maggie wakes in a hospital bed to see the Chief of Staff sitting beside her. She recognizes the room as one in the military field hospital on the periphery of the Moogh Zone. The Chief of Staff is calm, patiently waiting. His presence is like that of a block of hardwood, coated in khaki-colored paint and speckled with shiny metal objects.

"How are you feeling?" he asks, as Maggie comes too.

"I am okay; I guess," Maggie sits up and looks around the green tent room. "Why am I here?"

"You passed out. We ran some tests. You're fine" he says, succinctly.

"But why in a military hospital?" Maggie rubs her upper arm, suddenly feeling cold. True to form, the Chief of Staff avoids answering a question by remaining silent.

"And what about those people?" Maggie asks, somberly.

"They're gone now. It was very hard for you to see that."

"Yes. It was." Maggie looks at her hands, melancholically. "I don't understand why they did that."

"People who believe silly things, do silly things," says the Chief of Staff. "Someone set a rumor that the Moogh wouldn't come back unless there was a sacrifice. That's a very cruel thing to put into people's minds."

Maggie looks at the Chief of Staff mistrustfully and says, militantly, "That's funny coming from the guy who wants to kill the Moogh. Or should I say, neutralize the Moogh threat?"

"I wanted to talk to you about that."

"What's there to say?"

"Off the record, Maggie."

"Maybe."

"I can't speak to you on the record."

Maggie wonders why she is so combative when normally she'd be calm. She's angry, that's the heart of it. Angry that bad things are happening to her and that she's vulnerable to powerful people. “Okay then, off the record.”

The Chief of Staff leans forward, speaking conspiratorially, “I have always viewed the Moogh as a wake-up call,” he says. “Like the canary in the coal mine warning human civilization to change its ways. And given that the US military is the most ecologically destructive organization on the planet, you’d think that we might do some waking up ourselves.”

“How are you going with that?” Maggie asks, curtly.

“We’re just getting started. You see, the plan is not to kill the Moogh, Maggie. The plan is to capture it.”

“Why?”

“We’ve undertaken super-computer modeling of the Moogh’s metabolism, and the results are stunning. It is incredibly energy efficient. Plus, the algae in its fur is over 90% efficient at converting sunlight into lipids. If we can replicate the Moogh’s biochemistry, we could develop machines that never need refueling. Plus, we could manufacture low-cost fuels for our military. You’d like that.”

“I’d like that?” Maggie is surprised at the soldier’s audacity.

“Biofuel, Maggie. Sustainable, solar powered, carbon-neutral biofuel.”

“Will you stop waging war and killing people, too?” she asks, defiantly.

The Chief of Staff chuckles, “We’re not that enlightened.”

“Will you get rid of your nine thousand nuclear weapons?”

“Not until everyone else does,” he says defensively.

“And why are you telling me this?”

The Chief of Staff rests back in his chair and adjusts his hat on his lap. Then he stands. “There is a queue of people outside waiting to see you. You get well soon. That’s an order.”

"Why are you telling me this?" asks Maggie, concerned that he is leaving.

The Chief of Staff doesn't reply. Instead, he departs the room. The door closes behind him, and Maggie is left alone in the green tent room. She looks at her hands despondently. A few seconds pass and there is a knock on the door. It is a firm but gentle sound made from the knuckle of a woman's hand. The door opens a little way, and Maggie sees Maxine wearing a concerned look.

"Oh, baby look at you," Maxine coos. She moves over to the bed and smothers Maggie's face in her bosom. Maggie goes still, not feeling enough love to contribute to the hug. After what seems an age of smothering, Maxine sits on the chair where the Chief of Staff had been.

"If there's any good news to come from this," she says, "it's that this will never happen again."

"Really? How so?"

"The people that burnt themselves were part of a group of Adherents who sympathize with Moogh Underground," Maxine tells her. "The ringleader convinced the immolators that they could bring the Moogh back by self-sacrificing. Fortunately, the dominant narrative about the Moogh isn't created by Moogh Underground; it is created by moogh.org, a team of intelligent empaths.

"What happened?"

"We have excoriated the infectious meme, and the Adherents are at peace again.

Maggie feels her stomach churn. "What does that even mean?"

"The bad guys have disappeared?"

"Disappeared?" It's that word again, last used as a threat against Perrin. Something about a bullet and a barrel of bubonic acid.

"Now, when are we going to see the beautiful Maggie Tarp back on deck?" says Maxine, moving right along.

"I'm ready to go, really," says Maggie, "I probably just

need to get the okay from the medical people."

"How about I get that underway for you now?"

"Okay."

"Your annoying co-worker is outside. Shall I send him in?"

"Yeah, I guess."

Maxine kisses Maggie on the forehead and says, "See you soon, beautiful." She departs and Maggie waits a few seconds before the sound of male knuckles raps an aggravated tune on the door.

"Hi Paprika," says Perrin, "can I come in?"

Perrin enters the room sheepishly. He takes a seat next to her bed and looks at her silently for a while.

"Are you alright?" Maggie asks, concerned that Perrin has failed to make any glib remarks.

"I'm okay; I guess," he says. "What about you?"

"A bit worn out. And sad."

"I want to apologize," says Perrin, abruptly.

"For what?"

"For not being there when you needed me."

Maggie waves her hand, "You weren't to know what was going to happen. You did get the story, didn't you?"

"I did."

"Is it in?"

"Yep."

"More Moogh deaths?"

"Yep." Perrin looks at the floor, chewing his lip.

"Do you say anything intelligent and caring to help the public understand what might have motivated all that misguided passion?"

"Nope."

"So, what did you say?"

"I said the Moogh drove them to it."

"You're a shithead. You know that, don't you?" asks Maggie, flatly.

"I know that," says Perrin, looking at the floor and

nodding slowly.

"Well, that's a start, I guess."

"We're going to need a boat," he says, directly.

"A boat?"

"There's no Moogh stories unless we are around the Moogh, right? The Moogh is swimming towards the high seas. So, we need a boat."

Maggie is suddenly alert, remembering the Big General's remarks that they would neutralize the Moogh when it was in international waters. The new information from the Chief of Staff suggests that they will try to capture it there, instead.

"Where are we going to get a boat?" she asks.

"I have some ideas about that. So I need you to do something."

"Uhuh?"

"I need you to buy a weeks worth of frat-house food."

"What's that?"

"Baked beans. Loaves of bread. More baked beans. Tinned stuff. And I'll meet you on the jetty with a boat."

"Why would I buy bad food and not buy good food instead."

"We're not going on a cruise, Paprika, we're going on a mission. Dress and eat accordingly."

"Dress accordingly?" asks Maggie, chuckling. "That's interesting coming from you."

A Stolen Yacht

As requested, later that day, Maggie waits on the jetty with a pile of bags loaded with tinned food and loaves of bread. In the distance, the volcano continues to pour ash into the air, and the whole sky is tinged orange.

A sleek sailing yacht pulls up to the jetty under motor power, and Maggie is surprised to see Perrin behind the wheel, expertly guiding the vessel alongside. It is a 40-foot single master, a wooden boat with classical lines. It has a white painted hull and varnished wood paneling on the coach-house. The decking is pale gray with neat lines of black caulk between the planks.

Perrin throws the propeller into reverse, and the vessel comes to a halt alongside the jetty. "Quick, jump aboard."

"What a great boat," Maggie says as she hands over the first of the bags. "Where did you get it from?"

"Stole it. Come on. Come on."

"You stole the boat?" Maggie asks, astonished.

"Quick! Quick! You can discipline me once we're underway."

Maggie hands over the rest of the bags and steps over the stanchion rail. She stands on the deck, watching Perrin as he returns to the wheel and maneuvers the boat away from the jetty.

"Stow that shit down below, Cabin Girl," he snaps.

"Bugger you!" says Maggie, defiantly, pushing her knuckles onto her hips. She wants to do exactly the opposite, but can't figure out how to do that without looking stupid. So instead, she throws some scorn in Perrin's direction. "I can't believe you stole a boat," she says with the haughtiest voice she can muster.

"And I can't believe you are arguing with the Captain. Come on. Snap. Snap. Let's get ship-shape."

Maggie swears under her breath and then collects the bags and carries them down the companionway. Below decks, she is pleased to see that the boat is as well presented as above. In the low light, the surfaces have the sheen of polished wood. There are brass fittings where they ought to be, and even the fabrics on the seat covers and curtains meet her approval. Everything is clean and in working order.

"Hmm," she says to herself as she looks around the galley for somewhere to stow the bags. The boat has come as a bit of a surprise. She lifts a hatch on one of the surfaces and sees a fridge packed with ice, wine and beer.

Once out of the harbor, Perrin turns the yacht in the direction of the distant smudge on the horizon that is the US warship. Then he lashes a piece of sheet rope to the wheel and steps expertly forward and hauls the main sail up.

Maggie comes back on deck and looks around. "How long until we get there?"

"That's about twelve miles away," Perrin says, stepping down from the deck into the cockpit.

"So. How long?"

Perrin adjusts the main sail. "Well, we're doing about five knots in this breeze, and the Moogh is swimming about three knots away from us. Then there's tidal currents to consider. And we don't know whether this wind will hold, increase, decrease or change direction."

Maggie looks perplexed, "So how long till we get there?" she repeats.

"F**ck knows. It's unknowable, and it's a boat trip. So relax, drink some wine. You're on a yacht now. And a damn nice yacht, too, if you ask me."

"Yes," says Maggie, "you have stolen a nice yacht. You'll have fond memories of her when you serve out your term in a prison cell."

"Not like that hasn't happened once already," Perrin chortles. "That would cheer you right up, wouldn't it?"

"Momentarily."

"Hey, you want to get me a beer?"

"No." Maggie says this as if it were obvious.

"Let me ask that another way."

"Why don't you try that?"

"Paprika, would you be so kind as to grab me a beer as I am a bit tied up right now, what with steering the boat and all."

"You know what, Perrin, I am feeling quite generous right now."

When Maggie returns to the cockpit with the drinks, she sees that Perrin has set out some cushions on the lee side of the cockpit. He directs Maggie to sit there, and she makes herself comfortable and pours a glass of wine.

"This is quite nice," she says, surprised at what the motion of a sailing boat trip can do to one's sensibilities.

"You look good on a yacht," says Perrin, grinning.

"So what do I do?" Maggie asks.

"Just relax. We'll have you to the Moogh in due course."

Maggie nestles back against the lovely fabric and closes her eyes, enjoying the warmth of the sun on her face.

She seems to slumber for some time before she speaks. "Perrin," she asks without opening her eyes.

"Yes, Ma'am."

"Why do you call me Paprika?"

"Didn't you used to write under a pen name?"

"How do you know that?"

"I looked up all your stories."

"I used to write under the name Marequa Taard."

Perrin starts chuckling. "Where did that name come from?"

"She was like my alter-ego when I was younger."

"And what happened?"

"I figured that if I were going to be successful I'd be better off using my real name. Even though it is Maggie Tarp."

"So there you have it," Perrin says. "Marequa, Paprika."

"And what about Tart?"

"Hmmm." Perrin thinks it through, wary that whatever he says could come back to bite him.

"You're working really hard on that aren't you?" Maggie grins like she were having a happy dream.

Perrin holds his tongue and waits patiently. After a little while, he sees the signs of her nodding off to sleep. Maggie's head lolls slightly to one side, and her fingers uncurl from around the stem of the wine glass.

"It's just wishful thinking, really," Perrin says quietly to himself. He leans over and takes the glass away from her, careful not to wake her. He flicks the glass empty over the side and places it where it in a safe spot.

As Maggie slumbers, she looks like a portrait by a master painter. Her mousy brown curls tangle around her shoulders and her delicate fabrics are crisp and bright. Her petite, desirous face seems to glow as it is warmed by the sun.

From his steering position, Perrin glances down at her periodically, feeling sense of pride that he has the privilege of watching over Maggie Tarp as she sleeps. He is also conscious that this is the first time that he has been able to enjoy her presence without the accompaniment of her prickly critique of his way of being.

Maxine's Net Worth

An hour passes with no noise but the sound of the waves lapping the hull and the breeze occasionally snapping the sails. Perrin is in a meditative state. He's still standing, holding the wheel, gently swaying with the motion of the boat, hypnotized by the horizon. Periodically, he grimaces, as one of his unhealed wounds plays up.

Then spontaneously, he breaks the silence by inhaling deeply and releasing a long, drawn-out sigh. An involuntary muttered comment follows, "Oh, Mooghy, it's so good to be away from the org."

Maggie rises from her slumber, blinking her eyes open and looking around. Everything is as it was when she nodded off. Immaculate. Exquisite. A calm sea. A beautiful sailing boat. She is even happy to see Perrin there, a competent seaman, watching over her as she rests.

"What?"

"We've escaped moogh.org and the murderous Latvian lesbian."

"Don't be ridiculous. Maxine's not Latvian. She's from Russia. And you're just saying that because you fell foul of her," Maggie is dismissive as she adjusts herself in her seat and reaches for the wine bottle and the empty glass. "Did I drink all that glass?" she asks, confused.

"You were going pretty hard," Perrin says, smirking. "And there's other reasons."

Maggie pops the cork and slowly pours the fluid, careful to balance against the gentle rolling of the yacht. "Like what?"

"Like the fact that Maximum Hydrazine was in the room while her henchmen tortured me," Perrin says. He enunciates these words so calmly that they seem to blend into the light breeze. The words are ephemeral, there and not there at the same time.

Maggie stops pouring the wine, stuck by Perrin's

comment. If he had been speaking glibly, she would have waved him off. But he said this with such an expressionless tone that she is left wondering how it was even possible.

Then she is reminded that Perrin's experiential base is much broader than hers. She thinks that in the past few weeks that she had been bumbling around the Moogh Zone, Perrin has had a very different existence. In that time he had been arrested in an airport, done jail time, got yelled at by the Dim Director, witnessed a murder, suffered torture and a near death experience as well. Imagine if his whole adult life was like that, she thinks. No wonder relating to him is so hard.

Maggie replaces the cork and lowers the bottle to the deck. She sips her wine, thinking it through, feeling inadequate to comment. Glancing up at Perrin, she sees that he is observing her in the way that a master does a student who needs constant tuition. She is also reminded of many occasions when Maxine seemed to hide another side, and this corresponds with Perrin's statement. She finds herself nodding, a grave look on her face.

Perrin continues, "And, I'm saying that because I know a lot more about moogh.org now. Oh, boy, what a racket." He starts chuckling to himself and shaking his head, knowingly.

"How would you know that?" Maggie asks.

"Well, after they tortured me, I was bed-ridden for days with nothing to do but get angry or get to know my torturers better. See if I couldn't understand why the so-called pacifist organization was so damn violent. I found out. Do you know how much she's worth?"

"I don't even know what that means."

"If Maxine sold everything she owns, her properties, artworks and shares in Moogh Capital Incorporated and added that to the cash in her bank accounts, how much money would she have?"

"I don't know," shrugs Maggie.

"Three point four billion at the last count three months ago."

"How would you even know that?" Maggie snaps.

"From documents on record. You've just got to dig."

"That's ridiculous."

"Damn right it's ridiculous. She's making a killing out of the Moogh and with a UN mandate to boot. Talk about a savvy business woman."

"I mean your story is ridiculous," says Maggie, defensively, "Moogh.org is a not-for-profit, and she's just the Chair."

"And moogh.org commissions Moogh Capital to run everything. And Moogh Capital takes a margin on everything. Everything. The catering, the security, all the goods and services that are sold inside the Peace Park. And Maxine is the major shareholder of Moogh Capital. And that Moogh Coin, I mean bugger me!" Perrin shakes his head, astounded by what he has learned.

"What about it?" asks Maggie, sitting up, engaged now. The Moogh Coin had always seemed an oddity to her. It facilitated people to gamble on the Moogh, and this was quite out of character with the moog.org stated empathies.

"In the last financial quarter," says Perrin, "it netted her - personally - six point four million a week. Do you get that? A week."

"Bullshit!"

"I've got the data on my laptop, down below."

"She told me she was struggling," says Maggie, deflated.

"Struggling!" Perrin bellows a laugh. "Struggling to stow all the cash, maybe."

"I can't believe that you just found this information."

"I had a week in bed recuperating with nothing to do but heal and try to understand why moogh.org is so free to murder people. I just followed the money, and there she was: Maxine Oxidizer Inside. Meat grinder inside, more-like."

Maggie adjusts herself in her seat, and a glum look comes over her face.

"And do you know how many Moogh Deaths there have been?"

"Don't get me started on that!" Maggie snaps.

"Let me rephrase. Do you know how many people have been killed by other people in the vicinity of the Moogh."

"About forty."

"Forty-three on record. But there is many more."

Maggie looks up at Perrin, seeking more information.

"I checked the security log that was written after Thomas Watt got shot. There is no record of the death. And that's odd, right? Even after I put the news in the paper. But that event was given a classification, EF5. You can check back through the logs and see another two hundred EF5 events. Then you cross reference with news stories about missing people and chatter on blogs about people going inside the Moogh Fence without permission and bingo. Almost every time there is an EF5 event, a few days later, someone reports a missing person. It's incredible."

"So what are you saying?"

"Moogh.org is killing people. About one a week for the past four years."

"Two hundred and forty Moogh Deaths," Maggie hears herself say.

She glances up and sees Perrin giving her that knowing look. She turns her head away and observes the immaculate sailing boat. Then she realizes that she is sitting here, slightly drunk and completely calm for a reason. Perrin needed to tell her what he had found. Perrin needed to warn her. And how else could he get through to her except to chill her out in a place where he could have her undivided attention.

She looks back to Perrin and sees that he has gone back to his silent observation of the horizon.

"Thanks," she says, quietly.

"It's okay."

Maggie looks glumly at her drink, thinking through all this new information. She glances over her shoulder to see the land in the distance, slowly receding, and she too feels relieved to be away from the order and control of the Moogh Zone. She wonders whether this might also be the reason for the Moogh deciding to swim out to sea. Even the Moogh is trying to get away from Maxine.

"So I think we just stand off at this distance until I've had a snooze," says Perrin, interrupting Maggie's thoughts.

"What? A snooze?"

"Yeah. I'm going down for a few hours, then your turn. So I'll leave you with the wheel. Is that alright with you, Cabin Girl?"

"I don't know what a Cabin Girl is."

Perrin starts making a chuckling noise as he steps aside from the wheel. "Come over here. I'll show you how to steer the boat."

Maggie's Dream

After a four-hour sleep, Perrin returns to the deck. It is dark, and Maggie is patiently standing at the wheel, guiding the boat in the direction of the distant lights of the American warship.

"A few more hours," he tells her, "then we'll be in the high seas. You should get some sleep. It might be a long night."

Maggie does as instructed. She moves down below and looks for a comfortable spot. There is a small cabin with a triangular bunk at the very front of the boat. It is beautifully appointed with quality linen. She crawls into the bunk and instantly feels that day's effort catch up with her. The last thing she remembers is reaching for the switch on the light and the cabin being plunged into darkness. Next thing, she tumbles into a dream.

> At first the darkness is warm and embracing but this changes to something brooding and discontented. The darkness is chilling as something cold rises from the depths. In the distance, an orange glow, flickering, the caldera of an erupting volcano coming closer. As it nears, it is the young woman on fire, pushing a little spurt of flame out of her mouth with every breath. Then an image of Thomas Watt 'single gunshot wound to head' lying dead at the Moogh's feet. The pacifist and the dead. Perrin's swollen eye was swiveling back and forth as he lay trussed with rope and tape in the boot lined with black plastic. Hovering over all, an evil presence, Maximum Hydrazine, the master puppeteer getting richer and more powerful at six million dollars a week. And there are hundreds of people running in a panic on a beach, scrambling to get traction in the soft sand, funneled into the narrow track that leads up the dune. The tide recedes, going out further and faster than it ever has and then it

returns, a rushing, rising wall of water that roars across the wet sand. The ground is trembling and people are shouting, "Paprika! Paprika! Paprika!" The tsunami water rushes in, roaring and foaming and Maggie is there, paralyzed with fear, unable to move, helpless in front of this watery monster rolling towards her. The first wave dumps on her, and she vanishes inside the swirling foam. She comes to the surface, gasping for air and she hears that noise again, calling "Paprika! Paprika!"

Maggie wakes, disoriented. It is dark, and she sits up and braces herself. Her world is moving and sloshing around all over the place. It is comfy, warm and dry, but outside, there is madness. There is the terrific noise of howling wind and the sound of waves bouncing off the hull. The yacht surges up and down and heels awkwardly to one side. For a moment, Maggie is afraid then she is alerted to one noise rising above all the others. It is Perrin calling her name, "Paprika! Paprika!"

Maggie steps off the bunk and promptly falls over because the boat is rolling and pitching violently. She grabs hold of a handrail on the ceiling and staggers towards the hatch. It is dark outside, and Perrin is illuminated just by the small light on the compass.

"There's a storm," he shouts at her over the noise of the wind and waves slapping against the hull. "Get the lifejackets."

"Where are they?"

"Under the seat at the table!"

Maggie returns to the cabin. It is dimly lit, and the vessel is moving so much that she is barely able to see or stand. The sensation of falling is overwhelming, and she staggers clumsily towards the table. Finally, she lifts the lid and retrieves two life jackets. She staggers back to the hatch and calls out to Perrin, who is engrossed in keeping the yacht on course.

"I've got it," Maggie shouts from the hatchway.

"Throw one here and put the other one on," Perrin yells back.

"Are we going to be okay?" Maggie asks, concerned.

"Yeah. You stay right there. I might need you."

Perrin reaches forward for the lifejacket and Maggie can see that he is in pain. The unhealed wounds are being torn by the physical effort of staying upright and keeping the yacht on course against the storm.

Maggie struggles into the lifejacket, pulling the zip up under her chin. She looks around the insides of the yacht, feeling as though this tiny chamber is the full extent of her universe for the time being.

She watches Perrin with admiration as he braves the cold wind and the constant white spray that lashes through the air. Far from suffering, he seems to enjoy the intensity of the experience. She observes that while he is fighting the elements, he doesn't make any noise, and she has an insight into the speed at which his mind works. If he is not having to sail a boat in a storm, then he needs something to keep that brain occupied, such as a string of quick-witted utterances.

There is a shuddering noise coming from above, and the vessel seems to vibrate. Perrin looks up and seems to wince. The wind picks up, and the boat heels to one side.

"Paprika!" Perrin shouts.

"I'm here!" she calls back from her spot inside the hatchway.

"I'm going to reef the mainsail!"

"I don't know what that means!"

"I need you to take the wheel for a while!"

"Oh, really?"

"Up the front, there's more of this wet weather gear! Put it on under your lifejacket! Do that now!" Perrin's instructions are clear and firm, and Maggie follows them without question.

When she gets back on deck, Perrin moves her behind the wheel and stands behind her with his hands on her wrists. He explains what she has to do and then he leaps up on the deck next to the mast. He yells out instructions, and she does her best to figure it out. It is chaotic, and the sensation of controlling a boat with the pressure of the wind and the moving ocean is at once exciting and scary.

Perrin staggers around on the deck. In one moment he seems to be in command of the job and the next, he is clinging to the boom, holding on for dear life. He threads pieces of rope through holes in the sail and eventually cranks the sail back up with a winch handle. By the time he returns to the cockpit and takes over the wheel from Maggie, he is panting, and his face is flushed and wet with salt spray.

"You okay?" he asks.

"I'm alright."

"You should go down below and get warm," he says.

"What about you?"

"I'm okay. Maybe you could make some coffee."

"Really? Is that even possible?"

"You'll have to tie the coffee pot to the stove."

"What a bizarre concept," says Maggie and she gingerly moves below.

Eventually, she returns to the cockpit holding two cups of steaming coffee. She hands one to Perrin and with the other, she slumps down into a corner and tucks herself in tight.

Perrin observes her as she moves gingerly back into the position with the cushions and gets herself comfortable. "Jump back in your little hidey-hole there, Paprika Tart," he tells her, grinning.

"Do you think the Moogh will be able to swim in this weather?" she asks over her coffee cup.

"Nothing about that thing would surprise me," Perrin says. He blows on his cup, and a waft of steam rolls up and disappears in the wind.

"Do you know where it is?" Maggie tries to look up over the gunwale, but the cold wind forces her back into the pocket of warm air that occupies the corner where she sits. She looks up at Perrin and notices that he has adopted that look that he has when he is concentrating on something distant or faint. The last time she saw it was on the night of the immolations.

"What the hell is that?" asks Perrin gazing into the distance. A long way off, there is an orange light that seems to be rapidly expanding in size. Maggie detects concern in his voice, and she rises from her seat and sees an object hurtling towards them at great speed.

Instantly it becomes clear, and Perrin screams out, "Get down!"

At that moment, a helicopter, flying just meters above the water surface screams overhead, passing the yacht by just a few meters. The noise is terrific - a hellish, bellowing whine - as it hurtles through the air. The heat and smell of the exhaust pulses across the yacht's deck. The chopper instantly turns and comes around in a full circle around the yacht.

The pilot is flying without a helmet, and the chopper is so close that Maggie can identify him. "That's the Chief of Staff," she gasps.

"What the hell is he doing," Perrin shouts above the roar of the chopper as it circles around them.

The Chief of Staff, raises his fingers to his forehead, almost like he was signing off. Then he turns the helicopter towards the warship, and there is an audible change in the chopper engines as he picks up speed.

The helicopter drives at full speed and scores a direct hit on the warship conning tower. There is a tremendous noise – BWAM! – as it penetrates the superstructure and then explodes, blasting an orange gout of flame punctuated with pieces of shrapnel and blown-apart sailors.

"Holy Pup!" shouts Perrin. His coffee cup clatters on the deck as he pulls the wheel around and hauls in the mainsheet to pick up more speed. "Hold on," he shouts. "We've getting out of here!"

Maggie raises her head to look around. She clutches the gunwale as the movement of the yacht changes, and she loses balance.

A siren wails on the warship as the bridge calls action stations. Seamen can be seen running to action stations around the deck. A machine gun opens fire. The noise of the cartridges discharging is matched with high-pitch whistle as lead flies overhead and thousands of meters per second.

The deck of the yacht is illuminated by green streaks as the tracer bullets illustrate the path of the machine gun fire. Orange sparks fly out of the warship's funnel as the engines go to full steam ahead. The huge ship accelerates and heels as the course changes, and she heads out to sea. In so doing, she brings her after-guns to bear and these fire off with a massive percussive pulse, which make Perrin and Maggie wince at every shot.

As the distance opens up between the yacht and the warship, Perrin starts to feel like they may have escaped with their lives but then there is the sound of something hard and fast impacting the yacht.

As the warship pulls away, a machine gunner strafes the yacht and bullets slam into the hull and pass through the sails. Pandemonium breaks loose, and the immaculate woodwork is shattered by lead projectiles and splinters of teak shrapnel.

Perrin grabs Maggie by the jacket and drags her to the floor of the cockpit and lies on top of her as the wood splinters and shards of shattered bullets whizz all around. The meat grinder lasts only seconds, but it seems like hours.

Once the madness subsides, Perrin raises his head to see

Maggie's face illuminated faintly by the compass light.

"You okay?"

Maggie nods numbly. Perrin rises to look around. He moves down below for a few minutes then returns to the deck holding a yellow plastic bucket.

"Paprika, grab this," he says, as he moves back to the cockpit and ignites the main engine with the turn of a key.

"A bucket? What for."

"Get down below and start bailing."

"Bailing?"

Maggie feels a pang of anxiety, and she leans towards the hatch and looks down into the cabin. What was previously an immaculate chamber of exquisite fabrics and polished panels now a ruined place. Chunks of wood and splinters are strewn all over the place. The beautiful cushions have been blasted into clumps of stuffing. And most concerning of all, the polished wood decking is now sloshed with water. Where did that water come from?

Maggie looks around to see Perrin is leaning over the side, looking for something. He straightens back up and accelerates the engine.

She notices that Perrin has a furtive look on his face, almost like the first time that she has seen him out of his comfort zone. Even when he had only one eye functioning in the boot of the car he had a composure that he now lacks. She hears herself asking the question again. "Are we sinking?"

Perrin runs one hand over his face, and the other holds the wheel. He adjusts the wheel to keep the boat on course such that the bullet holes below the waterline are kept out of the water. It seems like an age before he looks up and sees Maggie in the hatchway, waiting for an answer.

"Not if you start bailing," he snaps. "Come on Cabin Girl. Go! Go! Go!"

Keep Bailing

The bailing goes on throughout the night. Bucket after bucket is filled in the cabin, lifted up the stairs of the hatchway and then tipped into the cockpit where it drains through little drain holes in the corner.

Perrin moves down below periodically to check on progress and the navigation chart. He sets the vessel on a course to the nearest piece of land and then takes turns with Maggie to either steer the vessel or bail water with the bucket.

They manage to stabilize the water at about mid-shin depth. The water slops around madly inside the cabin as the yacht continues to battle the heavy seas. Perrin returns to the cockpit and clears debris from the drain.

"Will we sink?" Maggie asks as she empties the umpteenth bucket of water.

"That depends."

She goes below and responds when she comes back up again. "On what?"

"On whether we can get the water out of the boat as fast as it is coming."

Maggie goes down below for more water and returns to the hatchway. She is soaked through and her hair has come loose and wraps around her face. She fumbles with the bucket and half of it sloshes onto her wet weather jacket and back down into the cabin.

Perrin starts laughing and wincing, holding his chest, "That's so funny."

Maggie struggles to pull her hair out of her face. When she can finally see, she points over the rail. "We ought to be sailing that way. Where the Moogh is."

"We could do that. Then we'd sink within the hour."

"Okay."

"Alternatively, we stay on this tack with the bullet holes out of the water and we might reach shore before we sink.

Which would you prefer?"

"Huh!" Maggie goes down below with the bucket again.

When she returns, the exhaustion has set right in. She sits in the hatchway, breathing deeply and slowly, holding on to the bucket full of water and wood splinters. The horizon seems to rise and fall as the yacht straddles the waves. Maggie wonders how the Moogh is faring in the big seas. It seems totally strange that her pursuit of the Moogh should have taken her to this place. A place where all that exists is the rolling and pitching of the sky and sea, and a constant sense of imminent death from drowning.

On the Beach

It is mid-morning when the yacht closes on the shore, a long stretch of sandy beach. When the keel hits the sand, Perrin dashes onto the deck and pushes the boom so that the sail catches the wind and the vessel swings around the right way. The yacht heels over with the hull coming to rest on the sand. Perrin peers over the seaward side and shouts excitedly, "The holes are out of the water!"

Excited, he goes looking for Maggie and finds her down below, slumped over the bucket, exhausted. The inside of the boat is a mess. Bullet holes pepper the varnished wooden paneling. Water has sloshed everywhere, and anything that is not screwed down now swirls in the knee deep flotsam. There are cushions, books and wood splinters lapping around Maggie's legs.

She's still wearing the orange wet weather gear over the delicate fabrics of her silk shirt and dress and looks like a doll washed up on the beach. She has salt encrusted on her forehead, and her hair looks like fishing line tangled up in seaweed.

She looks up deliriously as Perrin clambers down below and raises the lid of the navigation table. "What happened?" she asks, weakly.

"We're not sinking anymore." He moves up the ladder with a metal ruler. Halfway, he pauses, "Are you okay?"

Maggie swills around in the bilgewater, struggling to form words, "I'm just umm…"

"You can stop bailing now." Perrin departs with the ruler and jumps off the boat, onto the beach.

Maggie slowly rises and looks around the ruined yacht and then at the bucket in her hand. Strange, she thinks, that the boat is not moving around anymore. Cautiously, she moves to the hatch and puts her head out to see a calm sea on one side and a sandy beach on the other. She moves back down below and when she returns she has a pair of binoculars with her. She steps onto the upper-most section of deck and peers into the distance.

"Hey Paprika, come and check this out," Perrin calls out. He is standing in waist deep water with a ruler placed up against the side of the hull.

Maggie peers over the side, and he looks up at her. "Look at that. They shot us with a fifty caliber machine gun. The mean bastards."

"I have to go to the Moogh," Maggie says, instinctively.

"We need to fix the holes in the boat, first."

"I have to go to the Moogh."

"You ought to see yourself," Perrin says, beaming a smile, "you look like you've been shipwrecked."

"I have to go to the Moogh."

"You are not thinking straight, Paprika. There are other priorities."

"My name is not f**king Paprika!" she bellows.

Perrin watches as she moves away from the side of the vessel and out of view. "Oh, man," he mumbles. He walks around the grounded yacht and steps on board.

Maggie is down below, sloshing around in the flotsam. She has retrieved the gear bag containing all the journalistic technology and is emptying the contents onto a dry surface in the galley. She handles the wireless camera preparing for an ill-thought through journey.

Perrin seats himself in the hatchway, passively watching her. "You don't want to go there."

"I have to be with the Moogh."

"Do you remember the sequence of events that led us to getting washed up on this beach, Paprika?"

"Yeah. You stole a yacht, and I bought beans." She shoves the camera back into the bag. "I have to know what happens next."

"I can tell you that. You go wandering into the desert, and that's the end of you. It's simple." Perrin taps the side of his head for emphasis.

"You should come with me."

"I have to fix the boat. Then I can come with you."

"How long will that take?"

"I don't know, a day maybe. I haven't figured it out yet."

"Then you fix the boat, and I'll go to the Moogh."

"You are not going to do that." Perrin raises his foot, barring her path through the hatchway.

The effect is instantaneous. Maggie sloshes through the water towards him and points a finger at his face in the same way that Maxine did when Perrin first arrived in the Moogh Zone.

Instantly flushed with fury, Maggie's teeth bare and maximum hydrazine pumps through her veins, "Don't f**k with me, Perrin Speer!" she bellows. But she is unable to sustain the anger, and it fades quickly and she slumps forward, staring at the wet decking.

When she looks up, she sees Perrin smiling at her, wryly. It disturbs her that he is so patient. How will she ever get what she wants with Perrin Speer standing in her way?

"So when I take a photo, it uploads to the thing, right?" she says.

"To the server. That's right."

"So why don't you fix the boat and check the thing on the laptop. You'll be able to see where I am from the … the…"

"From the GPS metadata. You're crazy; you know that?"

"And you're a shithead."

Perrin laughs, "You know what journalists call that?"

"What?"

"A good team."

Maggie chuckles wearily and looks back to the bag in her hands.

"I'll tell you what we should do," Perrin says. "We're both worn out. Let's sleep for four hours. Then I'll make sure you have everything you need to undertake your madcap mission. Like water, for example. Deal?"

Maggie lowers the gear bag, nodding her head, defeated. She sloshes through the flotsam and slumps face down on the bunk where she had her troubled dream.

Perrin sighs deeply, relieved to have headed off her suicide mission. He steps into the cockpit and lies down on the seat in such a manner that he can see if Maggie comes up the hatch. He raises his arm and takes off his watch. He squints wearily at the numbers as he sets the alarm, allowing four hours sleep. Then he flips up the hood of his wet weather jacket and lodges the watch next to his ear.

He rests his head back against the wooden seat. His mind whirs as he thinks through all the possible means by which he can persuade Maggie to give up her obsession to go into a desert on behalf of the Moogh. Sleep rolls over him like a green wave.

In the Desert

Below decks, Maggie lies face down on her bunk, her eyes wide open. She can't sleep, and she sits up, feeling as if the boat is still moving around, even though it is parked solidly on the beach. She is exhausted, worn out, and yet feeling as though there is something more to be done before she can sleep. What is that? It's the Moogh, of course.

She slips off the bunk and very quietly makes her way to the galley to the bag with the wireless camera. She lifts the bag and cautiously approaches the ladder. Peering into the cockpit, she sees Perrin flaked out, making that annoying noise that he made on the plane when he slept. He's been up for days, but he's not full of scotch so who knows whether he'll wake easily or not.

Maggie steps back into the boat and takes stock of her situation. She is wearing canvas flat soled shoes, an orange dress and silk shirt. Over this is the wet weather jacket. She has a bag with a spy camera and a quest to see the Moogh. What else does she need? A silent exit.

She moves back to her bunk and activates the handle that holds down the hatch. The hatch rises, dripping salty water into her face. She lifts herself onto the deck then reaches down to retrieve the bag. Then she makes her way as quietly as possible to the part of the boat that is closest to the sand and jumps down onto the beach. She slings the bag over her shoulder, looks along the beach, first in one direction and then in the other.

She thinks through the sequence of events, the storm, the bullets, Perrin changing tack to keep the bullet holes out of the water. Feeling confident with her navigation calculation, she sets out along the beach, adjusting the bag full of gear against her back, trying to find where it is least uncomfortable.

The sand is soft, and every footfall sinks an inch driving

up the energy require to keep pace. Maggie manages to walk about a kilometer along the beach before the energy drains out of her, and she collapses on the sand. She rolls onto her back and lays there breathing heavily, looking up at the sky.

Her clothes are drenched, and she hasn't eaten properly in days. She's sleep deprived and feels as though she's been rolling around on the yacht for a week, one minute floating, next minute sinking. Above her, the clouds swirl around like clothes in a tumble dryer.

In her delirium, Maggie is reminded of the night that she drank wine with Novell. How long ago was that? Another lifetime or this one? What was it about that day that stands out? She told Novell that she had touched the Moogh once. She had never shared that before. And now she feels an emptiness. A yearning for company. A yearning to be with the Moogh and to know that it is safe.

She struggles to her feet but has not enough energy to move. So she bends over, her hands resting on her knees looking at the sand. She slowly raises her head to look along the beach. One long deep breath and she lurches forward and gets her momentum up again.

Every step forward is labored and seeming to go nowhere. She walks like this for hours, deliriously, watching the sandy beach move slowly underneath her. Periodically, she stumbles, falls on her face, rolls over and catches her breath, helplessly, like an upside down turtle. Each time this happens, she is driven to continue, and she staggers to her feet.

It is mid-afternoon when she comes to a halt. She doesn't fall this time, just props herself up with her hands on her knees staring at the sand. She repeatedly blinks, feeling giddy and yet coming awake. She runs her palm over her face and squints at the sand in front of her.

Does she see right? Is that a set of giant footprints in the sand leading from the water's edge, up the beach and into the dunes.

"The Moogh has footprints?" the dimwitted Director seems to be right there. Maggie's heart starts to pound and her tiredness wafts away.

She stumbles forwards, following the footprints. Even though the sand is at a steep angle, she powers up the dune. When she reaches the crest, she observes the footprints continue down the other side.

Maggie lowers the heavy bag to take stock of her surroundings. She scans the horizon and sees nothing more than the desert, punctuated by the occasional shrub or cluster of rocks and boulders. Behind her a cobalt sea, flat and calm despite last night's storm. The long sandy beach that seems to go on forever in both directions. In the distance, from where she came, a little dot that is the bullet holed yacht that Perrin stole.

Maggie draws a deep breath. Then she slings the bag over her shoulder and makes her way down the dune, following the Moogh's footprints.

Sand Storm

Maggie walks for hours following the prints, sometimes feeling like she was going in the right direction at other times, unsure. As the day drags on, the feeling of exhaustion creeps over her again, and she realizes how poorly equipped she is for the journey.

Her mouth is parched, and Perrin's remark, 'like water for example' keeps playing over and over in her mind. She wonders whether he has fixed the boat and has checked on her progress through the data uploaded from the camera. There's a point, she thinks, she has yet to take any photos.

Maggie retrieves the camera and switches it on. As she walks, she pans it around, looking for something significant to photograph. But there is nothing standing out. She halts and focuses on the horizon up ahead. The camera has a powerful zoom lens, and she adjusts it and her spirits rise immeasurably when she catches sight of an object in the distance. It is small because it is distant. It appears to be moving and even has a recognizable gait of the Moogh ambling.

She takes a photo and watches the little green light twinkle, and the image magically wafts away into the ether and onto the server. Then she puts away the camera and picks up her pace in pursuit of the Moogh.

It is late afternoon by the time she comes alongside. The Moogh is ambling at a fast pace deeper and deeper into the desert.

"Moogh!" she cries out, hoping to divert it from its path. But the Moogh neither stops not turns to look at her. She steps in front of it and puts her hands out, imploring it to stop, "You can't go there."

But her pleas have no impact and the Moogh just keeps ambling into the desert. Maggie halts right in front of the Moogh's tracks and announces, with uncharacteristic authority, "I'm not moving."

But the Moogh just walks around her and keeps on its way. She runs in front of it again and this time she grabs hold of its foot and tries to physically arrest its movement. Without looking down, the Moogh raises its leg and Maggie grips tight.

"You're going nowhere," she says, but all this does is to halt the Moogh for a few moments. It raises its leg, and Maggie loses grip and slips onto the sand. Feeling defeated, Maggie falls in next to the Moogh, and they amble together through the sand.

She feels exhaustion coming on again and takes stock of her situation. She is wearing comfortable shoes; that's a good thing. She is also wearing a cotton frock and an orange wet weather jacket. These clothes have been drenched in seawater and have dried and covered in crystallized salt. They have been torn and smeared with oil from bailing out the yacht.

And Maggie's hair, now there's a story. Normally, her hair is fine and light, and it rhythmically bounces as she walks. Now, however, it looks like it has surrendered to gravity altogether. At one point during the day she had bundled it into a manageable clump with a hair band. But since then half of it has fallen out, and the salty, oily mop hangs limp and forlorn around her cheeks. The occasional strand finds its way into her mouth, and she pulls it aside, conscious of the taste of the yacht's bilge.

The Moogh, on the other hand, seems to have benefitted from all the salt water and desert air. It's fur catches the evening sun and seems to impart a golden glow.

Maggie looks wearily ahead into the hazy distance, noticing the horizon has taken on a peculiar form. She watches intrigued as the distant sky becomes orange in color, and a long, thin cloud comes into relief. It seems to unfold in slow motion, but, in fact, it is moving towards them at speed.

"Mooghy? Do you see this?" Maggie asks. She takes hold

of the Moogh's fur and watches as the monstrous ball of orange dust, a desert sandstorm, approaches. "It's going to get really dusty, really soon."

The Moogh says nothing in reply, as could be expected, so she responds to her own statement, putting on a rendition of Perrin's accent, "Batten down the hatches, Cabin Girl." She giggles and replies in an accentuated version of her own voice, "Aye Aye, Captain."

The onset of the dust storm comes more quickly than she expects, and the wind picks up, dry and irritating to the airways. Maggie is inadequately prepared for this new challenge and she halts to adjust her clothing. She pulls the zip on her jacket all the way to the top and buries her nose and mouth inside of the fleecy collar and breathes through that.

Next, she seeks to put her hair all inside the hair band but reaching to the top of her head is difficult because the weatherproof jacket restricts her movement. She struggles with this normally simple task as the wind howls around her. Finally, her hair is arrested, and she can breathe without asphyxiation.

She starts to walk ahead but realizes that she's lost in the cloud of swirling dust. The sun is low on the horizon, and the dust filters out most of the what light there is. In an instant, Maggie finds herself alone, deprived of her senses. She can't hear anything other than the howl of the wind. She can't see anything other than the veil of orange that seems the same color no matter which way she looks. All she can smell is the aroma of the yacht, the scent of salt and diesel fumes imbued in the fleecy collar.

She feels a wave of panic wash over her, as though she were going to suffocate in this lonely, invisible place. And alone too, because she has lost the Moogh.

"Moogh!" she shouts out and immediately gets a mouthful of sand. She splutters, squinting, trying to determine where Moogh is, but she's totally disoriented.

She has the same harrowing feeling of having lost the Moogh as she had on the beach that very morning, and she starts to get anxious.

Maggie calls out again, this time with her hand over her mouth to prevent the ingress of sand. But the noise is muffled as a result, "Moogh! Moogh!" she cries desperately. "Where are you?"

The wind whirls around and forces her to close her eyes from the grit. The environment overwhelms her, and she falls to the ground on all fours, gasping. Closer to the ground, the wind is less intense, and she can make out the faint impression of a big footprint.

Maggie traces her hand around the print, determining which way the toes are pointing. Then she starts to shuffle on all fours in that direction. One footprint leads to another, and then another, and she realizes that she has only a little time left before the sun sets. If that happens, then even these faint reminders of the Moogh's passing will disappear from view.

Maggie runs ahead a dozen paces then falls to the ground again and checks for Moogh prints. She finds them and then rushes forwards again. After repeating this four or five times, she notices that the prints are getting noticeably fresher, having suffered less exposure to the dust storm. Her heart starts racing as she feels like she has a win. She lines up three footprints and then makes a mad sprint in their direction.

She get about ten paces and runs smack bang into the Moogh's back. The impact knocks her off her feet, and she sits there, stunned, disoriented. Through the swirling mist, she sees the faint shadow of the Moogh. It has stopped and thrown a pose.

The relief is overwhelming, and Maggie finds herself in stiches of laughter. She gulps a lungful of desert dust, nearly chokes, and this only serves to make her laugh harder still.

She comes to her senses, realizing that the Moogh could start ambling at any moment and disappear into the darkening dust again. So she crawls on all fours, coughing and laughing right up to the Moogh. She rests against its leg, using it as both a windbreak and a backrest.

"Oh, my," she says. "What a strange life I lead."

Maggie stays in this position as the sand storm passes, and the sun goes down, casting the desert into darkness. She hugs herself, keeping warm inside her jacket while the Moogh continues its pose. Weary and hungry and thirsty and yet calm and happy, Maggie nods off to sleep.

The Moogh Coos

Maggie wakes to find herself lying on the sand in the desert. The clouds have gone and above her is the glow of the Milky Way tinged orange by the remnants of the volcanic ash still circling high in the atmosphere.

The Moogh is a few meters away, and it is moving in a manner that she has not seen before. It seems to be cycling its poses, throwing one, holding it for a few seconds and then throwing another. It seems to be a ritual, like it was reliving his life over. Maggie is intrigued. She pulls the camera from the bag and takes a series of photos.

Then, even stranger, the Moogh stops posing and lays down on the sand. It tucks its right arm under its chin and then with the left hand it starts to excavate the sand, digging a hole.

When the hole is sufficiently deep, the Moogh adjusts its position to bring its right hand close to its face. Then it starts to make a noise – the first noise that it has ever made. It starts to coo over the object in its hand. Then the tears start to flow. It opens its right hand and allows its tears to saturate the object inside.

Maggie is stunned. At last, she can see the thing over which there has been such speculation. The object is the size of a small egg. It is dull bronze in color and with the Moogh tears it glistens in the starlight.

With one long sniff, the Moogh gently places the object inside the hole that it has excavated. It drags some sand over the hole and then starts to wriggle in its place, dislodging the sand.

Maggie steps back, stunned, watching the Moogh crying and wriggling. The more it wriggles, the deeper it's body sinks into the sand. She walks around snapping pictures. After about fifteen minutes of this, there is little left of the Moogh above the sand apart from its cute face with wispy hair and baby eyes.

The Moogh raises these doleful eyes and for the first time, it looks directly at a human being – at Maggie. Maggie squats on her haunches and looks into the Moogh's eyes. As she watches, she fancies that she sees the Moogh raise a smile in her direction. Does the Moogh know that she has been its longest serving human companion?

The Moogh's eyes flicker closed and then it makes its final movement. It reaches out its arm and drags a pile of sand across its face. Then it extends its arm so that it is underneath the sand, and the Moogh disappears altogether.

Maggie is stunned. She stands there staring at the ground astonished that the Moogh has simply disappeared from view. She can make out the pattern of the disturbed sand but the Moogh its self is completely underneath. It's as though it wanted no one but her to know its final resting place.

Maggie sits on the sand, forlorn, exhausted. She tucks the camera into the bag and looks around. She is in the middle of an empty landscape illuminated by stars billions of light years away. She is all alone, in a foreign country and her dearest companion in the world, a mysterious 11-foot tall primate of unknown origins, has just buried itself in front of her eyes.

Maggie sighs, trying to make sense of it all.

"Oh, Mooghy," she says.

A Sapling

Maggie wakes to the low sun warming her skin. The memory of the night's events come back to her, and she remembers the strange ritual and the passing of the Moogh. Last night, it all made sense. But this morning, it is all out of place. She stands and looks over the Moogh's grave, feeling a swirl of many types of emotions.

She becomes short of breath, confused and not knowing how to respond. She feels sadness that the Moogh will never amble or pose again and yet relieved that it is at peace. Next she feels guilty for so quickly letting the Moogh go. Then she is questioning whether it is dead or just sleeping.

She prods her finger into the sand, feeling the Moogh's fur just beneath. "Mooghy?" she asks. But the Moogh doesn't respond. She prods again and is reminded of when she first touched the Moogh. It doesn't feel like that anymore.

Maggie sits down and places her hand over the patch of sand that covers the Moogh's head. She inhales deeply, feeling her emotional flux resolve itself and her breathing adopt a normal pattern. What comes out of all of this is a calmness and an okay-ness with the way things have turned out. After all, what could she have done to change events?

She is not sad by this new turn of events, nor particularly surprised. There is something poetic and logical about the Moogh having swum across the sea and then expired in a desert. Maybe the Moogh simply wanted some peace and quiet in its last days. Maybe it had simply had enough of all the people prodding it and interfering with its affairs. There are so many possible explanations, Maggie thinks. The Moogh's passing has left so many questions unanswered. But is that not the way that the Moogh had lived its entire life, from the day of the Genesis Amble?

Maggie is not sad that the Moogh has died, nor happy

about it. She is simply content in the knowledge that the Moogh is where the Moogh wanted to be. And to this end, things are as they always had been. The physical manifestation of the Moogh will return to Gaia from where it came. But the spirit of the Moogh and the things it left behind - including thousands of acres of Moogh daisies and billions of fat bees - these things remain.

Observing the disturbed sand with the Moogh beneath, Maggie feels a sense of completion, of having seen the story through to its end. Plus, she is aware of her privileged status of having been present for the Moogh's entire life. Maggie had known the Moogh from the from the day of the Genesis Amble. Back then, she was a backpacker, unsure of how her life would turn out. She was the lucky girl who happened to be in the right place at the right time, and the only one on the bus kept her head while all the others started screaming, falling into Mooghmania.

Maggie looks down at her ground, nodding her head, knowingly. It is then that she notices something very strange and yet not strange at all.

From the sand, in the place where the tear moistened hole had been dug, a small plant has grown. It has a thin stem and two leaves that point towards the sun. Not only has this plant grown in just the past few hours, but it is also visibly growing now!

Maggie blinks as she stares at the plant. Does she see this right? The growth rate is so rapid that its expansion is observable just by watching for a few moments.

Then the insight comes to her, and she gasps aloud. One of the big Moogh questions answered. What was the Moogh carrying in its right hand all this time? It was a seed. A seed for a Moogh Plant! That the Moogh is dead is no surprise. It seems so natural, like the ruination of the egg to allow the chick inside to escape.

She lies down in the sand so that her face is close to the little plant and watches it intently, seeing how the growth erupts from a pale green tip. As it reaches higher, the existing parts of the plant become darker and fuller, the leaves growing in size. Then a little lump appears on either side of the tip, and this take on the form of two new leaves. And it seems that the more leaf there is, the more sun it can absorb and the faster it can grow.

The growing tip is now above her head, so Maggie sits up to be at eye level with it again. She pushes out a finger and touches the plant. It feels just like a plant should, which is no surprise, but as her finger makes contact, the tip seems to turn towards her, as if to reward her affection.

Then underneath the first set of leaves, which are now level with her chest, tiny red beads appear and Maggie peers at them, fascinated.

"Oh my giddy aunt," Maggie says, suddenly overwhelmed by what she is seeing. While she has become accustomed to the oddities of the Moogh, this new Mooghery has her in quite a bind.

She stands and takes a step back and instantly jumps in surprise! There is something behind her! She turns quickly and sees, towering over her, another freshly grown plant, this one a bit taller than the other. Its large oval leaves seem to be faces looking at her.

"What is going on?" she says aloud, stunned by it all. She turns back to the first sapling and sees that the little red dots under the leaves are now a cluster of delicious looking berries.

Having been without food and water for days, Maggie is instantly intrigued by this new turn of events. She touches the berries, and the cluster seems to come away from the tree and she holds them in her hands.

She sniffs the berries. They smell sweet and aromatic, so she pops the berries in her mouth and crushes them between her teeth. There is the sensation of a burst of fluid and the most awesome taste in the world and an instant sensation of satisfaction.

"I have eaten the fruit of the Moogh tree," Maggie tells the Moogh tree. "And I feel good for it." She retrieves the camera from her bag and paces around the cluster of fast growing saplings that now forms a little grove around the resting place of the Moogh. She positions her camera in a particular place and takes a photo every five seconds to mark the growth of the foliage.

By now, the tree is big enough to offer shade and Maggie puts away the camera and rests against the trunk, watching the foliage grow and wondering whether it will grow forever. The berries regenerate and Maggie picks and eats these.

On the underside of the leaves, condensation forms and a tiny trickle of moisture works its way down the stem where it pools in a hollow at the base of the branch. Maggie puts her mouth to this tiny pool and slurps and receives enough water to make the effort worthwhile.

Suddenly, Maggie is exceptionally happy. She jumps in her place and claps her hands together. "Of course, the Moogh was carrying the seed for the Moogh Tree." It all makes so much sense.

Maggie runs around the perimeter of the Moogh Grove observing its extent and the growing banquet of food as dozens of different types of fruit sprout from the stem.

Then something very odd catches her eye. Away in the distance there is a figure approaching. It is a man, but a man covered in some strange devices. Maggie steps back into the grove and hides behind a trunk and looks out cautiously. After a few minutes, she recognizes the man's gait, and she steps out and calls, "Perrin! Over here," and starts waving happily.

Perrin Finally Gets It

Perrin approaches looking sunburnt and tired, but with eyes full of life. He halts a few meters away from the grove and leans back to take stock of its extent. He starts shaking his head and then makes his characteristic chuckle.

"What on Earth are you wearing?" Maggie asks, astonished. She observes Perrin nodding at her, knowingly, almost as if Maggie is to blame for something.

Strapped to Perrin's back is one of the big solar panels off the yacht. He wears a bulging backpack across his belly with a tangle of wires running out of it. One set of wires snakes around his back to the solar panel and the other set connects to the laptop computer strung around his neck on a piece of cord. In his hand is a GPS receiver. He looks like a confused techno-turtle.

"See if you can work it out," says Perrin, wearing that look of the school teacher overseeing a student who needs extra tuition. Maggie paces around him, checking out all the pieces variously connected.

"What's in the backpack on your front?"

"A very heavy battery and an inverter to keep the laptop going."

"Hmm. So the solar panel charges the battery," she says, thinking it through.

"That's about right. It's how I was able to track you."

"Are you angry with me?" Maggie asks, coyly.

"I was yesterday," Perrin tells her. "Then I got caught in a sandstorm."

"Hmm."

"But that was yesterday. Today I am pleased to see you."

"Hey, are you hungry?" Maggie asks, excitedly. "Come around here."

Perrin unbuckles himself from his solar powered computer gear and lowers it to the ground. He follows her into the grove. There is a clearing in the center where the disturbed sand that covers the Moogh is visible.

"Is that the Moogh?"

Maggie nods silently.

"How are you with that?"

"I am okay, I guess. It hasn't sunk in yet. But look over here. The Moogh Tree wants to feed us." She points to the various types of fruit growing on the branches. "We could have a picnic."

"I've got some bread and beans in the bag," Perrin says.

Shortly, the picnic is underway. Many types of Moogh Fruit, white bread, tinned beans and a bottle of warm, white wine drunk directly from the bottle.

"Hey, what happened to the yacht?" Maggie asks, breaking from the feast.

"Oh, that yacht sank," says Perrin, plainly.

"That's sad. I liked that yacht."

"We'll just steal another one."

They start chuckling together, and Maggie says, "Well, there's that, I guess. How will we get back to the office?"

"There's a town a few kilometers that way. Maybe we can get a taxi from there to the city and then jump on a plane. We should do that soon."

"I guess."

Perrin falls silent, just staring at the sand. Maggie looks across to him, sensing the melancholy emanating from his direction. "Are you alright?"

"When I woke up, I'd slept for hours through the alarm. The tide was high, and the boat was taking water again. When I found you'd gone, I felt really bad that I'd not come with you."

"I snuck away, remember?"

"Yeah. I guess." Perrin nods gently, thinking it through. He picks up a handful of sand and lets it slip through his fingers. "I've done some soul-searching," he says. "I think that about ninety percent of the things that I have ever said to you have been mumbling gibberish."

"I'd go along with that," says Maggie, honestly.

They look at each other, almost like it is the first time that they have agreed on anything.

"I need to learn how to shut the f*** up," Perrin says.

"That depends what you want for your life."

"I want to be more like you. Caring about things."

"Then you really should learn to shut up."

"Any words of advice?"

"Well you have already started," says Maggie. She places her hand on his wrist, and he looks up at her. "So, keep going, I guess."

Write Up and Submit

After the picnic, Maggie and Perrin depart the Moogh Grove and walk in the direction of the nearest town.

As they walk, Maggie is thinking about the passing of the Moogh and the subsequent growth of the tree that turned into a grove. She has mixed emotions about it all. One part of her misses the Moogh badly, and another part is content that all is as it should be.

The walk to the town is laborious, and she stares deliriously at the sand and she trudges forward. If not for Perrin at her side setting the pace, she would probably fall face down on the ground like on the beach.

It takes most of the day to reach the town and then to locate somewhere to stay for the night. They find a small hotel and check in as the sun is going down.

In the dining room, they dump their gear next to a table and retrieve laptops, camera, and notepads and all the kit necessary to tell the world the story. One last push to write up the story and then they can sleep.

"You know this is the biggest story in the world right now," says Perrin, flicking through the photos that show the last hours of the Moogh and the rapid growth of the Moogh Tree. "There're a billion people around the world who would crawl over broken glass to read this story."

"I guess," Maggie says.

"Serious. The Moogh has been gone for days. Everyone will freaking out thinking that the US military has killed it or run off with it. This is the scoop of the Century and you are on point."

"I don't know if I'm up for that," Maggie says. She is sitting in front of the laptop, waiting for it to boot up, looking disheveled and worn out in her orange jacket.

"With your permission, I'll guide you."

"Sure, let's do it then eat and sleep."

"Okay, I want you to smash out two thousand words

that follow the events laid out in these pictures that I've selected. I'll write up a bunch of little stories for context and crop and edit the photos. Can you do that?"

"Yep." The computer is ready. She opens a new file and sits looking at the blank page, waiting for the muse. She takes off her wet weather jacket and lays it on the floor. Perrin starts to laugh.

"What's the matter?"

"You look like a mop. Do you want to take a shower first?"

Maggie shakes her head and starts typing.

She spends the evening writing up the story from when they departed the harbor on the yacht. They describe the storm, the helicopter crashing into the Navy ship. She describes the sandstorm and the Moogh doing a funny dance, burying the seed and shuffling itself into the sand. She describes the rapid growth of the Moogh Tree into the Moogh Grove. On that point, she strikes a firm agreement with Perrin to never disclose the location of the Moogh's body. In keeping his word, he deletes the file that contains the GPS co-ordinates.

Maggie types quickly and consistently for three hours. Once the final keystroke is made, she ceremoniously closes the lid of the computer and then seems to wither into her jacket.

"I'm so stuffed," Maggie says.

"You look it."

"Did you sort the flights and the business with customs?"

"Yep. With credit cards and cash, you can get anything in this country. We fly the day after tomorrow afternoon. I reckon we get there early and drink heavily."

"Okay. Can I sleep now?" Maggie crosses her arms in front of her on the table and slumps forward, her head cradled in the fabric of the wet weather jacket. In so doing, she dislodges the stack of paperwork that Perrin has unpacked from his bag and has been sorting into various piles. A vanilla folder falls from the table.

The noise brings Maggie to attention, and she leans over to look towards the file on the floor that has spilled its papers. On the cover of the file, she sees two words that are written in Perrin's distinctive scrawl: 'Operation Augmentation'. That's curious. She remembers that word being abused by the Dim Director. This piques her interest.

"I'll get it," says Perrin, leaning forward but this compels Maggie to get to it first. She leans forward and extends her hand towards the papers that have spilled from the file. She stops, maybe from exhaustion, or perhaps because there is something odd written on the sheet of paper lying face up, something that triggers an immediate curiosity. She stares at the words for a while, then turns her head just enough to see that Perrin has observed the situation and is wearing a distinctly guilty look.

"Perrin?" she asks.

"*Ahhh...*" he replies, cryptically, unable to complete the sentence.

"Is there something you want to tell me?" Maggie leans forward and shifts the document so that it is completely free of the folder.

"I was going to tell you about that?"

Maggie retrieves the folder and opens it on the table in front of her.

"*Uh-huh*? And then what happened?"

"Then this crazy guy flew a helicopter into a warship."

Maggie looks up at him, disbelieving. "What is this?" She raises the document, a printout of an email chain.

"It's a commissioning document," he tells her. He looks down at his hands. "I wanted you to know about it. I mean, I want you to know about it."

"Commissioning document? Commissioning who?"

"Commissioning me."

"To do what?"

"Augment you."

Maggie thinks back to the day they first met. The Dim Director had decided to 'felicitate her augment-alation'. Was Perrin in on this?

"So… I don't get this." Maggie looks back to the paper and tries to figure it out.

The document in front of her is an email chain. The first email is from: terry423@gmail.com and to hydrazine@moogh-capital.com. It reads: "Hi, Maxi. Copping a lot of flack from the board about Moogh Desk sales. One of my hacks (you remember Perrin?) has offered to spice things up a bit. Room for one more embed in moogh.org?"

Then, from: hydrazine@moogh-capital.com to terry423@gmail.com: "Hi, Teddy - Sure, send him down. Tell him to keep away from the crown jewels. BTW, capital raise coming up soon if you want some more stock…"

Maggie sits back and crosses her arms. She looks directly at Perrin and recites, "One of my hacks has offered… It was your idea?"

"Perrin looks down at his hands, bashfully. "I was out on the piss with Terry one night—"

"You're on first name terms with the Dim Director?" Maggie asks, astonished.

"Yeah. Anyway, he tells me that the board wants to shut down the Moogh desk because not enough people were reading your work."

"So you thought you'd just take over?"

"Well--"

"And what's with Maxi and Teddy?"

"So the Dim Director and Maxine had a thing one night and--"

"A thing?"

"Yeah, it was a cocktail party, and everyone got pretty messy."

Maggie raises her hand. "Hold on. You were there?"

"I guess."

"So you socialized with Maxine?"

"And Teddy."

"You set this whole thing up. And all this is going on over my head, and you never thought to tell me?"

"We didn't really have that kind of relationship until--"

Perrin is unable to finish. Maggie stands suddenly. Her face is flushed red with anger, embarrassment and sadness all at once.

"You know what Perrin Speer," she says, her voice wavering, "you can just go to Mars." She seems totally stunned as she moves away from the table and slowly walks out of the dining room, her head low.

Perrin slumps in his seat, shaking his head, wearily. "Oh, Mooghy," he sighs. Then he lifts the lid of the laptop and stares at the words that Maggie has written.

He sighs again and then turns his mind to his work. He thinks through all the actions that need to be taken to tell the world that the Moogh has died with a byline that reads: Maggie Tarp.

Dim Director Excited

Maggie sleeps a full thirty-six hours – a night, a full day and another night – before she finally wakes. She lies under the covers, thinking it all through. She's thinking about how her whole world has just been turned upside down.

The Moogh is gone and so is her sense of how simply the jigsaw pieces of her life fitted together. Once there was moogh.org and the Fractious News Network, two jigsaw pieces that didn't fit together, except that she was a piece that joined them together. But now there are all these new pieces, Teddy and Maxi and the 'thing' that they had. There is Moogh Capital that Teddy owns a bit of. And there's Perrin Speer at a cocktail party while Teddy and Maxi have the thing.

It's all too hard. Maggie pulls the blanket tighter against her face and sighs, "Oh, Mooghy."

Eventually, her curiosity gets the better of her and she craves to know how Fractious News has dealt with her story. She retrieves her smartphone and pulls it under the cover and checks her emails.

There is a message from the Dim Director. She doesn't read it, just notes its exuberant tone and the verbose praise for excellent efforts and the quality of her copy. He's added a link to the Fractious News website, and she clicks this.

The front page is dedicated to the Moogh. The headline reads: 'Another Moogh Death – this time it's the Moogh'. The lead story has had over ten million visits, and there are thousands of comments. For the most part, a vast outpouring of grief.

She does a quick scan of the leading media outlets and finds that they are all carrying Moogh stories, sourced from Fractious News.

"Kaching!" she says, sarcastically, imagining the

Fractious News Directors grubbily rubbing their hands together at the thought of all the extra revenue.

Maggie checks a variety of other journals and is surprised to see that even they lead with Moogh news. She shakes her head in astonishment. Even the Economist carries a Moogh story on its front page, about Moogh Coin. With the Moogh no longer ambling, the Moogh Coin currency was collapsing, and hundreds of millions of dollars were at risk of being wiped out.

She goes back to the Fractious News site and observes that the stories are surrounded by adverts for oil and motor cars and expensive shampoo with polypropylene micro-beads that choke turtles and fish. On seeing this, Maggie feels her stomach turn, and she closes the web page. Selling toxic shit, that's all her stories come down to. Nothing to do with truth or empathy or philosophy.

She turns back to the Dim Director's email and reads that he is recommending her for a promotion, pay rise and a seat on the company board. He has called a full board meeting and wants her there as soon as possible.

Under her blankets, Maggie thinks this through. Wondering what it all means. It strikes her as odd that just hours after she learns that her superiors and coworker socialize together, she too is being invited into the fold. How does that work?

As she thinks it through, she has her first realization of just how out of touch with all this she is. As the embedded journalist with moogh.org, she had been able to swan around in her bubble for years now, never understanding the nature of the relationships that allowed her to be there.

She is reminded of the comment in Maxine's email warning that Perrin should keep away from the crown jewels. What did that mean? Maybe the crown jewels were the Moogh Ninjas, and that is why Perrin was so horribly treated for breaking the story. Maybe that was Maxine's way of getting back at the Dim Director for... What?

Maggie forces another big sigh, "Oh, Mooghy!" then throws the blankets aside and sits on the side of the bed, psyching herself to start the day. Like it or not, understand it or not, the world lives outside her hotel room, and she has no choice but to engage with it.

Richard Corey

Perrin is waiting for her in the lobby when she finally makes her way downstairs. She slumps down in a seat across the coffee table from him.

"How are you going?" he asks, bashfully, hoping that she is not still sad or angry.

"I'm surprisingly calm. It is sort of like the lights have come on."

"Well, that's good," Perrin observes her carefully, noting a distinctive shift in her demeanor. It is like she has come out stronger from the ordeal, not weakened by it.

They take a taxi from the hotel and as the vehicle drives away, Maggie asks a question that had been playing on her mind.

"Why were you such an asshole to me on the first day. Haranguing me in the elevator?"

"I was hazing you."

"What does that mean."

"I was just testing to find out whether you were strong or weak."

"And what was I in your esteemed judgment?"

Perrin doesn't answer, and Maggie looks to him, seeing him holding that tracker dog look again. He is fixed on something that he has seen in the street.

"Driver. Pull up here. Quickly!"

The cab comes to a halt, and Perrin pulls the camera from its bag. He steps out of the cab and extends his hand to Maggie. She takes it and steps out with him. He points down the street, "There is some crazy ass shit happening over there. And to answer your question about my hazing, you are tough. Real tough. Just a bit wet behind the ears."

Perrin pulls out his wallet and hands the cab driver a bundle of cash then retrieves the rest of the bags from the taxi. "Come on Paprika. It's story time."

Half a block away, there is a well-dressed man, portly, in

his sixties. He wears a silk shirt open to his waist. He is ranting and pointing towards a house that has a tree growing out of its windows.

Maggie looks towards the guy shouting in the street and then to the tree growing out of his house. "That's the Moogh tree," she says, stunned.

The tree seems to have grown up through the middle of the house. It's branches have broken through the windows and are visibly growing taller and sprouting new leaves.

"What's going on here, mate?" Perrin asks.

"Who the hell are you?"

"I'm the journalist who is going to tell your story to the world."

"You want a story, here's my story. The tree is killing my business."

Perrin snaps a bunch of photos, and the old guy turns his attention to Maggie. She shies away, uninterested.

"I spend my life building houses and factories, and now the tree grows up and wrecks everything," the old man shouts, waving his hand towards the growing vegetation.

"How long's this been going on?" Perrin asks.

"It all starts last night. One minute no tree, then this. All over the town."

"There's more," Maggie asks, now intrigued.

"Damn right there's more! The people don't need my house because they live in the tree. And they don't work in my factory because they eat the fruit of the tree. So I try to kill the tree. But every bit I cut off, it grows back."

"That's awesome," says Maggie, instantly excited. "Can you show me?"

"This way," says the man.

"What's your name?" asks Perrin.

"Corey."

"Corey what?"

"Richard Corey."

The old guy walks down an alleyway, pointing out the

houses that he owns. The tree is consuming the whole street. Branches force their way through windows and doorframes. The walls are cracking where green tendrils force the bricks apart.

"You see this one," the old boy moans, pointing. This morning the tenant moved out." There is a dense forest growing around the house.

"Where did he move to?" Maggie asks.

"Says he is going to live in the damned tree."

Maggie looks back to the tree with a sense of wonderment. It is a dense mat of trunks and buttresses and in a few places that seem to be a natural entranceway inside.

"Thanks, Mr. Corey, that's all I need from you," says Maggie. She steps into the shade of the tree and acclimatizes to the lowered light. The air is cool and sweet smelling, unlike the aroma of the city tinged with petrol fumes and dust. The tree has formed a complex structure with multiple levels.

From an upper level, Maggie hears a voice sing out, "Hi, beautiful." She looks around for the source of the voice. It is a young guy wearing jeans and tee shirt. He's munching on a piece of fruit, more of which is growing in a cluster near him.

"How are you going?" Maggie retrieves her notepad.

"Couldn't be better."

"You look pretty comfortable up there."

"Home sweet home."

"Do you live here?" asks Maggie.

"I do now," says the guy.

"Where did you live before?"

"I used to rent off Richard Corey. I worked in his shitty factory."

"You are just going to live in the tree."

"Hell, yeah. It's got everything we need. Shelter, food, water."

"That can't be good for the economy," says Maggie.

"So what? What is the economy anyway? It's a tool to make people's life better. That's until psychopaths like Corey turn it into a means to control people for their own benefit. You don't need an economy in the tree."

"How big is this thing?" Maggie turns to take in the full extent of the tree.

"It's huge. It stretches for miles that way. And it gets bigger every day. A week from now it will have swallowed up all of Richard Corey's properties and then he'll be as rich as all the rest of us."

"And what will you do now that you don't have to work?"

"Have a real life," the guy says. He raises a guitar and strums a tune. "You know what I call this tree?"

"What?"

"About bloody time."

"That's a good name," says Maggie. She raises her notepad, draws a spiral in a circle, and makes a note of that comment.

Maggie moves through a naturally formed alleyways made by buttresses and branches. She sees Perrin, snapping photos.

"Pretty cool, huh?" she says excitedly.

"I like it."

"You know what I think?" asks Maggie. "It's like the way a caterpillar transforms into a butterfly. The tree is the Moogh in another form."

"Yeah, right," says Perrin, nodding, "I get it." He looks around in wonderment at the complexity and detail of the tree. "Hey, can you hear that?"

"What?" asks Maggie.

"It's a chopping noise. This way."

They move through the tree and see a man with an axed avidly chopping at a trunk. Perrin raises his camera and captures the scene as the man chops away at one stem while, unbeknown to him, another grows out of the ground behind him. It seems to grow in bursts with each pulse of new growth timing perfectly with the blows of the axe.

Eventually, the man breaks through the stem that he is chopping, and a piece of the Moogh Tree falls to the ground. He moves the axe head into his palm and wipes the juice and woodchips from the metal blade, nodding his head appreciatively. He turns to depart and is startled to see the new tree that has grown behind him. He looks up to see the foliage continuing to spread. Despondently, he lowers his axe and walks away.

"Oh, my," says Maggie. "This thing is unstoppable."

"I'm going to do an interview with that guy," says Perrin.

"Okay," says Maggie, "I am going to do an interview with the tree."

The Moogh Seeds

Back inside the tree, Maggie closes her eyes and becomes very still, sensing the living organism around her. She breathes deeply and listens to the noises. There is a sound of leaves rustling in the breeze above her head. She can hear a distant sound of a siren wailing, probably some alarm going off as the vegetation pulls the town's infrastructure apart. Then there is the distinctive sound of children laughing.

Opening her eyes, Maggie moves in the direction of the laughter and comes to a clearing where there is a man and three children. The man is hard at work rolling a strip of bark briskly between his hands, forming a rope. One of his ropes is already in use, tied to a bough, with three children are happily playing with it.

"Hi," says Maggie.

"Hey," the stranger replies, happily.

"What are you doing?"

"Having a life," he beams a big smile.

"What are you making?"

"Kids toys. Once, when I wanted to get the kids a present, I'd buy some non-biodegradable plastic rubbish that would probably end up in the river then the belly of some poor bloody turtle. Progress, they called that. This is much better. And look." The man holds up his hands to show off his palms. "Physical exercise. No need for a gym anymore."

"Do you live in the tree now?"

"Hell, yeah! It's the best real estate ever. You don't pay rent when you live in nature."

"What do you eat?"

"Look around," the man illustrates by waving his hand through the air. This thing is full of food."

Maggie observes a dozen forms of fruit grow from every part of the tree. There are fruits that look like apples, bananas, stone fruits, oranges, and peaches.

"See how fast the water flows when the dam breaks," the man says.

Maggie moves to another part of the tree and closes her eyes. After a few moments, she has a strong sense that she should head in one direction. It is as though the tree were calling her that way. She follows the instinct and soon comes to another clearing, surrounded by buttresses.

A beam of sunlight penetrates the canopy. Where the sunlight strikes, there is a small cluster of white flowers. They have a rich, fruity aroma, and Maggie leans close to smell them. She is reminded of the perfumery section of a department store, and she closes her eyes and inhales deeply through her nose. The aroma is intoxicating. She swoons, reminded of when she stepped out of Maxine's limousine after drinking vodka. The flowers emit a complex aroma, a mixture of scents including orange and frangipani.

When she opens her eyes, the cluster of flowers has grown and as she watches, it continues to expand. Like a fractal expanding, the flowers increase in number following a consistent pattern. In one moment, the cluster is the size of an apple, and then it expands and duplicates and is the size of a grapefruit. The flowers themselves are white with yellow insides and thin pink lines running down the sides.

As Maggie watches the flower cluster grow, she sees the pattern start to change and slowly, the bunch of flowers opens up and reveals an inner chamber. Inside this flowery hiding, space is an object that has the color of the seed that the Moogh planted.

The object is moving, albeit very slightly, and Maggie leans in close to see. The surface of the object is changing, transforming. The round object splits apart to form two pieces, and the two parts split into four. It is like watching a time-lapse sequence of a cell dividing.

Maggie moves her hand toward the object, to feel its

texture. As her fingertip connects with the smooth skin, the object detaches, and she instinctively catches it. When the object touches her hands it immediately falls apart, and Maggie stands there, startled, holding dozens of little Moogh Seeds.

"Oh my stars," she says aloud, genuinely surprised. She shifts the seeds around in her palm with her thumb, investigating them closely in the beam of sunlight. She chuckles mischievously as she thinks through her new fortune.

Then, the bunch of flowers starts to recede in the opposite manner to which they had grown. As she watches, the flowers disappear, and Maggie is left standing in the clearing unsure of whether any of this had actually happened. However, when she opens her hand and sees the seeds in her palm, she knows that it is not her mind playing tricks.

In normal life, this event may have seemed spooky or threatening, but Maggie lives in the world of the Moogh where such things are to be expected. The tree wanted her to have some seeds. Who was she to question that? She slides the seeds into her pocket. Then she goes to find Perrin and get him to the airport.

Long Live the Moogh

Back in London, Maggie's first task is to expand her wardrobe beyond a ruined dress, silk shirt, and wet weather jacket. After the shopping spree, she checks into her normal hotel and takes a long shower that includes a thorough hair wash. Dressed in her new clothes she feels less Maggie Tarp the average looking girl with freckles on her nose, covered in bilge grime and more like Maggie Tarp, the famous journalist.

She spends the rest of the day on the phone and reading the news as the story about the Moogh's demise continues to reverberate around the world. She calls her contacts in the Moogh Zone to hear how they are coping with the news. They are not coping well at all. The Moogh Zone is in turmoil.

On learning of the passing of the Moogh, the Adherents have rioted and burned down Moogh HQ and Tent City. With all the infrastructure that had supported them in smoldering ruins, they wander, lost and confused, the very fabric of their lives torn away.

The Moogh Coin currency has crashed, leaving millions of people around the world out of pocket. This outcome was an inevitable result of the Proof of Amble algorithm. Once the Moogh stopped ambling, the mechanism that kept the currency going was removed, and it could do nothing but tumble to its intrinsic value of zero. There had been fist fights and arrests outside Maxine's hotel as a baying crowd of ruined gamblers sought redress from the executives in charge of the Moogh Casino. As could be expected, the Level-4 Security were in the news for using excessive violence as they sought to control the riot.

Taking advantage of this chaos, Moogh Underground released a video that purportedly showed Satoshi Nakamoogho reading a statement. Just who is reading the statement is unclear as they are filmed wearing a Moogh

Mask. The statement describes how Moogh Underground plans to ensure that the entire moogh.org infrastructure is destroyed because the organization has deceived the public and were at odds with its pacifist charter. To kick this off, they have deployed hundreds of Non-Sticks, mainly unemployed youth, armed with spray cans. The Non-Sticks are tagging every bare surface with a new variant of the Moogh Emoticon. This one has neither vertical bars or circles for eyes, but crosses. The text sprayed under the new emoticon reads, "The Moogh is Dead. Long Live the Moogh."

Maggie puts a call through to Maxine on her smartphone, not expecting that she'll get a reply, and not certain that she wants one. Nonetheless, a reply comes through over the Moogh Messenger app. It is a short recorded video of Maxine. She looks frazzled and harassed. "Hi Maggie," she says, feigning normality, "Glad to hear that you are safe. Love to see you around the Moogh Zone sometime." Then Maxine looks away from the camera as there is a loud pounding at the door and a voice booms "Police! Open the door." Then the video cuts out.

"Oh, Mooghy," says Maggie, lowering her smartphone. She is reminded of just how much of a bubble she had created for herself. Two weeks ago, she could never have imagined seeing Maxine in anything other that perfect poise in command of the entire Mooghsphere. And look at her now.

After a few minutes, there is a newsflash. The reporter is being jostled by the crowd, and he says into his microphone, "Coming to you live from the Plush Hotel in the Moogh Zone."

The camera cuts to Maxine being escorted through the lobby by plainclothes police. Outside, an armed military unit are rounding up the Level-4 security and bundling them into the back of flatbed trucks.

"That's the end of moogh.org," says Maggie, stunned.

So much for the jigsaw puzzle so neatly arranged, each piece perfectly and sensibly connecting with the other and with Maggie being a vital intermediary piece.

She thinks back to the day that she broke free from the Adherents by doing the interview with Moogh Underground and sending it to Fractious News. And how quickly things have moved once that happened. Within weeks, she was on a salary with Fractious News and embedded with moogh.org. She had always taken this as a sign of her innate journalistic skill. But now she sees that she had accidentally exposed moogh.org for the fraud that it was and given Fractious News, Teddy, in particular, a weapon to hold over Maxine's head. Maggie had never been an interconnecting jigsaw piece. She had been redundant since the beginning.

She finds herself staring at a TV screen watching a compilation of violent incidents where Level-4 security are seen blasting and stomping and electrocuting the innocents who were simply demanding what had been offered them, the truth. Maggie sees that all her empathic stories about spirituality and Moogh philosophy had been aired for one reason only. This was to hide the truth that the executives had been milking the Moogh from the day the Declaration of the Adherents had been signed. Even the title of the document – Declaration of the Adherents – was a lie. She was an Adherent when it was signed, and yet it didn't contain her signature; Government Ministers signed it. And what did they ever care about the Amble and Pose of Moogh Destiny?

"Oh, Mooghy," Maggie sighs. She suddenly realizes that she is as complicit in the deception as Maxine and Teddy. But who is worse, them because they knew, or her because she was ignorant. The media, she remembers, doesn't tell you what to think, it tells you what to think about. And in that, she has been complicit.

As Maggie stares at the images of the Mooghsphere on

the TV, on her laptop and her smartphone, she becomes clear on three key points. First, Perrin Speer has been looking out for her right from the beginning, and he is the only source of truth in this whole sorry affair. Second, the entire edifice of business and the economy needs to change. All the elites and their smoke and mirrors and bullshit of public relations, and that horrid mess called business as usual, it all needs to be brought down and replaced. Replaced with something grand and clever that heals the Earth and fairly rewards all the people, not just a few. And third, she is free to say and do just what the Moogh she likes. After all, who is going to question the great Maggie Tarp while her name is attached to front page stories in journals and media outlets across the entire world. Just let them try!

Her meeting with the board of Fractious News is in a few hours time. That might be a good place to start telling some truths!

Cigarette Butts

On her way to meet the Directors of the Fractious News Network, Maggie checks her watch to find that she is a few minutes ahead of schedule. To kill some time she, sits on the wall that surrounds the little planter next to the steps into the Fractious News offices.

She looks down and sees dozens of cigarette butts lying on the pavement. She imagines the rain coming, the butts being washed into the gutter, into the river, out to sea where they will get eaten but not digested by fish and sea turtles and other endangered species. And the poor animals will have those cigarette butts fill up their guts until they can eat no more and then they will slowly starve to death.

Instinctively, Maggie leans forwards and picks up the smelly cigarette butts. She looks around and finds another piece of unsustainable, non-bio-degradable plastic, this one a chip packet in the planter. She puts the smelly butts in the chip packet and tucks this into the side of her handbag.

Damn this world that the humans have made, Maggie thinks, looking around at all the buildings, lights blazing in the middle of the day, sucking up vast amounts of electricity made by burning coal. All the planetary boundaries have been shot to pieces by global capitalism. The planet is dying, and only the Moogh made a stand to save it. Not only save it from decline but regenerate it, grow it back. But the Moogh is gone and in its place there is a tree.

Then she remembers that she has brought a bit of that Moogh tree with her, inside a little silk pouch in her handbag. A cheeky grin spreads across her face, and she looks around to check that she is not being watched.

She retrieves a stick from the planter and uses this to dig a shallow hole in the soil. Then she places one of the Moogh seeds inside the hole and covers it up.

"I am sorry that I don't have any Moogh tears for you

today," she tells the seed. She brushes her hands together and sits there, smugly wondering what might happen next.

Her thoughts turn to the meeting with the Directors. They want to promote her, give her more money and even sit on the board with them. Does she want that? In fact, now that the Moogh is gone, what does she want? As best she can figure it, she doesn't want much at all. She has been particularly carefree since the Moogh passed, and the lights came on about all of the Public Relations bullshit and corruption in the world. All she wants to do is whatever it is she feels like she wants to do at any given moment. It is a very liberating feeling.

Maggie picks up the stick and digs another hole. In this one, she places two of the Moogh Seeds. She chuckles as she covers the soil over. Then she fossicks in her bag for her small glass water bottle and sprinkles water on the disturbed soil. Standing next to the flower bed, she looks up at the tall building all the way to the top where she is now late for her meeting with the Directors.

It is just past nine in the morning on the ground floor of the offices of the Fractious News Network. In the palatial marble-lined lobby, Maggie waits for the elevator. She glances momentarily at her reflection in the shiny metal doors then looks away to contemplate what she has seen. It is an odd smile that she wears today, a sort of cheeky, troublesome smirk. If that look were put into words, it would say, "Don't mess with Maggie Tarp today."

The door of the elevator slides open, and Maggie steps back to allow the people inside to egress. She glances briefly at the people exiting and one person she recognizes.

"Hi, Perrin." She is surprised at how eager she is to see him. She has been ignoring his calls since she left him at the airport arrivals lounge the day before.

Perrin looks downbeat but upon seeing Maggie his face lights up. "Hey Maggie," he says enthusiastically. They hug in the lobby, and it's the first time that that has ever happened, and it is nice.

"Say, what are you doing now?" she asks.

"I was just going to get a coffee."

"You want to sit in on a meeting?"

"Sure. What is it?"

"I am going to nuke some Directors in the boardroom," Maggie's grin spreads across her face.

Perrin chuckles aloud, "I'd love to see that."

On the executive floor, Maggie advises the fittingly glamorous receptionist of her arrival and a few moments later, the Dim Director steps out of his office. The tall, imposing man with the squinty eyes is impeccably dressed. He seems to occupy a full one-quarter of the room.

"Maggie Tarp!" he announces, opening his arms as if to hug her. But Maggie puts her hands up and takes a step back.

"Okay," says the Dim Director. "What's he doing here?"

"Perrin's with me," says Maggie.

"I thought that you hated each other."

"Well there is something you don't know," says Maggie.

"Anyway. In we go." The Dim Director indicates towards the boardroom

Directors and Psychopaths

Maggie pushes open the door of the boardroom and ushers Perrin inside. It is opulent inside with lots of walnut and chandeliers and the Directors all seated around a massive table.

Maggie takes stock of them. They are old white folk, mostly men, but a few Dames in there, too. If you were to pawn all the jewelry attached to their pampered carcasses, you could feed an African nation for a year, Maggie thinks. The air reeks with entitlement and self-grandeur. It's stifling.

The Chairman stands and moves towards Maggie, with his hand outstretched. Maggie ignores him and moves to the head of the table and instantly starts talking. "I came here for good reason and not because you asked me."

The Chairman looks baffled. He glances at the Dim Director, who just shakes his head.

"If you wouldn't mind taking your seat," says Maggie, firmly, pointing in that direction. She waits until the Chairman sits and then continues.

"The first thing that I want to say to you overfed plutocrats is that your empire has hit its zenith, and it's on its way down. The water flows fast when the dam breaks. And the dam has most certainly broken."

"That's a rather forthright attitude," stutters the old Chairman.

"It's what you get when you pair an empath with an ambulance chaser," Maggie growls. "You ought to take some responsibility for that."

"What the devil is she talking about?" The Chairman turns to the Dim Director, who quite clearly doesn't know.

"What I am talking about - you stupid old walrus - is that this organization has all the characteristics of a clinically diagnosed psychopath. It purposefully exploits any person or thing to achieve its narrow objective of turning human and natural capital into money for a small number of individuals: the shareholders of the Fractious News Network. But this organization is not a person, it's a structure created by people. So the real psychopaths are the people who direct it. That's you lot, by the way."

"Bravo!" says Perrin, suddenly coming to his feet at the back of the room. A dozen stunned Governors look in his direction.

"Who the hell is that?" asks the Chairman.

"This is my associate Perrin Speer," says Maggie.

"You are making a terrible mistake, Maggie," says the Dim Director. "We were going to promote you and offer you a board position."

"I don't want to sit on your stupid psycho-board. And besides, your empire will soon be turned to lignin and cellulose."

"She's gone quite barking mad," says the Chairman, turning in his seat.

"Well, here is the interesting thing, Mr. Chairman," says Maggie. "You people who run the world so badly are the ones who are mad. But of course, you can't see your own madness. I mean, what sane person would knowingly destroy the living environment that supports them and do nothing to prevent its decline? Where do you stupid rich people think that you are going to go? Mars? A space station? The richest billionaire can't find seafood that isn't contaminated with flame retardants and plasticizers because of all the damned plastics in the ocean. Don't you get it? There is nowhere to go." Maggie taps the side of her head repeatedly to emphasize the point.

She continues, "Well, the planet has an answer to the mess that you lot have made. I have just returned from the Moogh Tree. It is growing rapidly and consuming every piece of stupid human infrastructure in its path. The Moogh Tree is going to wipe your rotten empire off the face of this Earth. And good luck trying to be psychopath living in a tree."

"I have absolutely no idea what she is talking about," says the Chairman.

Maggie has finished her rant, and she moves quickly through the room towards the door. She stops next to the Chairman, who wears a bamboozled look on his face.

"Oh, Mr. Chairman," Maggie says, opening her purse. "I always want people to get what they deserve." She slaps the chip packet full of cigarette butts in his hand. "Recycle that, you old git!"

Maggie storms out the door. Perrin follows close behind, a grin wrapped three times around his face.

In the lobby, waiting for the elevator, Maggie turns to see the Dim Director coming out of the board room. His face is red, and he bellows, "Maggie Tarp!"

The elevator doors slide open, and Maggie and Perrin step inside. Then the doors slide closed.

The Zenith of Civilization

"Bugger me, Maggie Tarp!" says Perrin, excitedly, once the elevator is descending. "That was amazing!"

"Pretty cool, huh?" she beams.

"Who taught you to do that?"

Maggie leans over and kisses Perrin on the cheek, "You did."

Perrin, places a finger where her lips had been. He has a look of stunned pleasure on his face. "Oh, really?"

"Thanks for teaching me," she says.

"I think you got that by osmosis."

"Hey, you want to get that coffee now?"

"I get to drink coffee with the great Maggie Tarp? Hell yeah! Can I get a photograph with you for my profile page?"

"Sure you can."

Perrin retrieves his smartphone and holds the camera out in front. They bring their heads together into the frame, and both pull suitably foolhardy grins. Maggie holds her fingers behind Perrin's head like rabbit ears.

"Click! "Got it. Thanks, Maggie."

"Call me Paprika. I like that. It's spicy."

"Okay, Paprika, I'll do that."

The elevator arrives at the lobby, and they walk across the marble floor together. Perrin steps ahead and pulls open the door, like a gentleman.

Outside, there is an air of apparent confusion. Something quite strange is happening on the street today.

"What the hell?" gasps Perrin. In the middle of the road, a tree has pushed its way up through the tarmac. Traffic has banked up on either side, and a police officer blows his whistle and waves the traffic past.

"That wasn't there when I arrived," Perrin says, perplexed.

"Funny that, huh?" Maggie laughs.

A cluster of branches are growing out from the base of a lamp post, causing the post to lean at an angle. An alarm bell starts ringing somewhere in the distance. And then another. And then a siren. Across the street, a window shatters and green vines extend from inside the building.

There is a crunching noise behind them. Maggie and Perrin turn to see a crack appear in the wall of the Fractious News office building. From within the crack, a green tendril appears and unfurls its leaves in the sunlight.

"Shot the duck!" says Perrin, stunned.

"The Moogh wasn't telling what we *were doing* wrong," Maggie tells him. "It was telling us what we *had done* wrong. And now it's putting it right."

"That's just ridiculous," Perrin says, watching as people pour onto the street as the tree consumes the office buildings.

"You know you are a pretty fine human being, after all, Perrin Speer," Maggie says, looking at him directly. "Right from the beginning, you were looking out for me."

"Oh, shucks, Maggie," Perrin blushes. "Everyone needs someone to cover their blindside."

Maggie opens her purse, "Put your hand out," she tells him.

Perrin does as instructed, and Maggie delivers into his palm a dozen Moogh Seeds. "I think that the Moogh would want you to have these. You want to help share the love?"

"Is that what I think it is?"

"What do you say it is?"

"An antidote to an unsustainable global economy."

"That's a good name for it. Do you want to plant some seeds of change?"

"Let's do this thing!" says Perrin, barely able to contain his excitement.

"We can plant them in the park down that way," Maggie says.

"Let's do it."

Maggie's eyes light up, and she walks down the stairs with Perrin by her side. Even without the Moogh, she is Moogh Happy.

End.

Dictionary of Moogh English

*See <u>General Notes</u> on pronunciation and spelling at document end.

255, 202, 0: The RGB colour code for Moogh Orange and also the official designation of an individual of unspecified gender. It is also colour of Lagom, just the right amount. Also, written as #FFCA00

Cavete Iniquitate Quae Est Ex Pace: The official motto of the Moogh Underground. It translates to: *Beware the Violence that Accompanies Peace*. It is in reference to moogh.org's heavy-handed management of the Moogh Zone.

Declaration of the Adherents: An international agreement of the United Nations General Assembly that established a Peace Park around the Moogh and the United Nations division of Moogh Affairs (UNMA).

Enlightenmoogh: The spiritual and ecological enlightenment that comes to people who spend time in the presence of the Moogh.

Entourage: All of the animals, particularly birds and insects that accompany the Moogh. These include honeybees that feed on the Moogh Daisies

Flybys: A derogatory term for temporary visitors to the Moogh Zone. So named after 'Fly by Night'; an untrustworthy or unreliable. Most are day-trippers who pay a hefty entrance fee.

Genesis Amble: The first observance by a human of the Moogh ambling on 3 January 2009. A Greek farmer who observed the Moogh ambling out of a cork forest first reported the Genesis Amble.

Lagom: A Viking word that describes an optimal portion, just the right amount. It is used as a greeting amongst the Adherents.

Level-4: Also known as Level-4 Moogh Security. This is the all-female security force recognizable for their stunning looks and immaculate mil-fash (military-fashion) uniforms. They wear white cork-leather cat suits with red straps and knee-high boots made of red cork leather. They are armed with Rhinox-25 stun guns, Walther pistols and carbon-steel trench knives. Level Four provide front line security around the Moogh and also accompany Maxine and other moogh.org executives for personal protection. They are colloquially referred to as 'stunners with stunners'.

moogh.org: The not-for-profit organization that was awarded the monopoly management rights over the Moogh. Moogh.org was founded and Chaired by Russian national, Maxine Slivkin.

Moogh Aficionados: A generic term for anyone who has a particular liking or interest in the Moogh.

Moogh App: The official social media platform for moog.org. It features the Moogle map, Moogh messenger, and Moogh Casino amongst other features.

Moogh Approach Protocol: The official protocol for approaching the Moogh. Level–4 Moogh Security are charged with ensuring that the protocol is properly followed and they have been known to use deadly force when the protocol is not followed.

Moogh Bees: A generic term for fat and healthy bees laden with pollen and full of nectar that have fed on the Moogh Daisies.

Moogh Capital: The privately owned company commissioned by moogh.org to provide goods and services necessary to manage the Moogh Zone. Maxine Slivkin founded Moogh Capital and she is the CEO and major shareholder.

Moogh Casino: An online gambling platform that uses Moogh Coin as currency. The statistical probabilities employed in the various bets leads to an average win for the house of 2.7%. Moogh Casino runs a net profit of millions a day.

Moogh Coin: A unique crypto-currency developed by Maxine Slivkin and used as the exclusive currency inside the Moogh Zone and for gambling in the Moogh Casino. At time of writing, Moogh Coin can be bought for US$230 each. Moogh Coin is generated through a complex algorithm known as Proof of Amble.

Moogh Coin Mining Rig: The mobile infrastructure used for creating (mining) Moogh Coin. The mining rig is comprised of thousands of Moogh ASIC computers inside a 40ft shipping container. A mobile refrigeration plant keeps the Moogh Coin mining cool. A 5 Megawatt gas turbine generator made from an adapted Mig jet engine mounted on a Russian armoured personnel carrier powers the Moogh Coin mining rig and the refrigeration unit. The generator is fuelled with Jet A1 fuel that is delivered daily in tankers.

Moogh Daisies: A generic term for the profusion of flowers that grows rapidly in the footprints of the Moogh. Moogh Daisies are the preferred food of bees which get fat off the nectar. It is possible to determine the path of the Moogh's amble, by following the Moogh Daisies. Huge clusters of these daisies grow where the Moogh has remained for a time during its pose

Moogh Emoticon: The official icon of the Moogh. It is formed from nine separate graphic elements representing each of the Moogh's nine poses. These elements include four curves (skull, chin, snout and mouth), two straight, vertical lines (eyes), two dots (nostrils) and one invisible element known only to senior executives of moogh.org.

Moogh English: A vocabulary developed by Maggie Tarp, Moogh English represents the official spelling and definitions regarding the Moogh.

Moogh Euphoria: An overwhelming sense of catharsis and that everything will be okay after. It is an emotional state often felt by people who spend time around the Moogh.

Moogh Fence: A protective barrier that surrounds the Moogh. Its form and distance from the Moogh varies according to terrain, perceived risk and other factors. Sometimes it is a rope held up by stakes driven in the ground. Other times it might be security personnel with arms interlinked, or even a picket fence.

Moogh Flag: The official flag of the Moogh that features a Moogh orange background with Moogh Emoticon and two crossed olive branches underneath. This design is referred to as 'Moogh and Cross Branches', a play on skull and cross bones, the infamous pirate flag, the Jolly Roger. Some variants of the Moogh Flag also incorporates the Latin words: '*cavete iniquitate* *quae est ex pace*' which translates to 'beware the violence that comes with peace'.

Moogh Flu: A hypothesized pandemic that is caused by viruses that move from Moogh to human that could kill off most of humanity. Whilst the World Health Organization dismisses the concept of Moogh Flu, it is a constant narrative of the Pentagon as a justification for 'neutralizing the Moogh threat' as they call it.

Moogh Happy: A combined sense of contentment and awareness that comes from having a disciplined, inquisitive mind and a positive affect. Not to be confused with Moogh Euphoria, which is something very different, indeed

Moogh Honey: The name given to honey that is made by bees that have collected nectar from Moogh Daisies. Moogh Honey is notoriously sweet and moreish and is used to make an intoxicating brew called Moogh Mead. Moogh Honey is also used medicinally for treating scratches, headaches, bee stings and even hangovers.

Moogh Infrastructure: All the structures, materiel and equipment that is required for the proper management of the Moogh and the people who interact with it. It includes the Moogh Fence, Moogh HQ bivouac complex, the Adherent's Tent City, the Moogh Radio truck and all the security apparatus. Moogh Infrastructure is continually being moved to maintain its appropriate proximity to the Moogh.

Mooghmania: A form of mass-hysteria in people around the Moogh characterised by uncontrollable screaming, and incidents of mass swooning. These psychological phenomena were commonplace in the early days when there was inadequate crowd control around the Moogh.

Mooglemap: A google map over-laid with Moogh Data including details of its ambling, posing and cry events. The Mooglemap is available through the Moogh App.

Mooghology: The official study of all things relating to the Moogh. Professionals in this field are referred to as Mooghologist.

Mooghologist: A professional scientist who studies the Moogh.

Moogh Mead: An intoxicating brew made from Moogh Honey. People who drink Moogh Mead report having intense, psychedelic dreams of a future utopian world in which there are no wars or fossil fuels and the oceans are not choked with non-biodegradable plastic. Production of Moogh Mead is strictly controlled by moogh.org although many people risk prosecution by operating their own beehives and distilleries.

Moogh Orange: The official colour of the Moogh with the colour code RGB 255, 202, 0 or #FFCA00. Moogh Orange is the official colour of 'unspecified gender' and the colour of Lagom, just the right amount.

Moogh Poses: The positions in which the Moogh holds is body when it is stationary. Shown below is the official Pose Number and colloquial name.

- Pose #1 I've forgotten something

- Pose #2 My shoelace is undone

- Pose #3 My head's on fire

- Pose #4 Stargazer

- Pose #5 Snail's pace

- Pose #6 Dr. Good

- Pose #7 Atten Hut!

- Pose #8 Swan Dive

- Pose #9 From whence it came

Moogh Salute: The highly symbolic, official greeting of Moogh Aficionados. It is performed with a two-fingered victory sign with the left hand, held horizontally across the chest with the knuckles adjacent to the heart. The 'V' shape is symbolic of victory and also of strength, representing the two fingers used to draw a bowstring. The 'V' emanating from the left to the right also represents the increase in natural capital and decrease in human demand for natural capital as per the Long Future Diagram. The proximity to the heart symbolizes honesty and love.

Mooghsphere: The totality of every thought, image, idea, product or conceptualization of the Moogh, around the world. The Mooghsphere did not exist prior to January 3, 2009. The Mooghsphere will persist until the last human forgets that the Moogh ever existed, which may never happen.

Moogh Security: The multiple agencies responsible for providing security for the Moogh and the people and infrastructure that surrounds it. Front line security is provided by moogh.org and incudes the feared Level-4.

United Nations Moogh Affairs provide security recognizable by their orange and blue helmets. There there are various police and military forces associated with the host country in which the Moogh is located. There is also a secret sniper-assassin team called Moogh Ninjas although they are rarely seen.

Moogh Tears: The lacrimal fluid that is 'drawn' from the Moogh during Moogh Tear events. Colloquially, these are referred to as crying, but Moogh English prohibits the word crying in this context. Moogh tears are drawn about once a month that these events prompt the Adherents and many other people around the world to cry at the same time.

Moogh Underground: The official protest movement against the institutions that surround the Moogh. A mysterious character called Satoshi Nakamoogho heads Moogh Underground. However, it has never been properly established if Nakamoogho is a male, female, 255, 202, 0 or a committee.

Moogh Zone: An internationally designated Peace Park in a 10-kilometer radius around the Moogh that is mandated by the United Nations Moogh Affairs, Declaration of the Adherents.

Moografitti: Graffiti that uses Moogh iconography. There are two main variants, that which conforms to moogh.org graphic design specifications and that which is adapted by the Moogh Underground.

Nonsticks: The colloquial name for disgruntled Adherents who leave the order without permission. Nonsticks make up the bulk of the membership of Moogh

Underground. They are called non-sticks because they no longer adhere.

NOTAM: A contraction of Notification to Approach the Moogh. A NOTAM is obtained via submitting a request through the Moogh App. It is the official mechanism that Moogh Security uses to determine who has been given permission to approach the Moogh. Maxine Slivkin approves all NOTAMs.

Pilgrimoogh: The journey of spiritual significance to visit the Moogh undertaken by millions of people.

Planetary Boundaries: An initiative of the Stockholm Resilience Centre, the Planetary Boundaries framework describes scientifically based levels of human perturbation of the Earth System upon which modern societies are dependent. The nine boundaries listed alphabetically include:

- Atmospheric aerosols
- Biogeochemical flows
- Biological diversity
- Climate Change
- Freshwater Use
- Land System Change
- Novel entities
- Ocean Acidification

• Stratospheric Ozone

Rhinox-25: An electric stun gun used by the ferocious Level-4 security. It is so named because it can stun a rhinoceros at 25 metres.

The Moogh: A bipedal, ape-like organism that mysteriously appeared in a field in France in January, 2009 and quickly developed a massive international following as a global messenger of peace and sustainability.

The Amble and Pose of Moogh Destiny: A concept held by Adherents and other Moogh aficionados about what the Moogh would do if humans did not interfere with it. It is also used to describe optimal management practices of the Moogh Zone.

United Nations Moogh Affairs: Also known as UNMA, this is the UN division established to co-ordinate the international management of the Moogh and the humans who interact with it. UNMA is a joint initiative of United Nations Environment Program (UNEP) and United Nations Economic and Social Co-operation Organization (UNESCO).

UNMA: United Nations Moogh Affairs. See above.

W61: A variable yield nuclear weapon developed by the US military. The bomb's explosive yield can be varied between 0.3 and 340 kilo tons. A kiloton is equivalent to a thousand tons of TNT high explosive and the bomb that destroyed Hiroshima had a yield of 15 kilotons - so the W61 going off with full yield is about 22 times more powerful. There are about 340 of these weapons built and they can be delivered by aeroplanes. There was some speculation that the Big General wanted to nuke the Moogh with a W61 to "neutralise the Moogh threat."

Made in United States
Troutdale, OR
11/03/2026

56852539R00125